Nature and the Crisis of Modernity

To my Mother,
and the memory of my Father

NATURE AND THE CRISIS OF MODERNITY

A CRITIQUE of CONTEMPORARY DISCOURSE on MANAGING the EARTH

Raymond A. Rogers

Montréal/New York
London

BLACK ROSE BOOKS No. X220
Hardcover ISBN 1-551640-05-5
Paperback ISBN 1-551640-14-7

Canadian Cataloguing in Publication Data

Rogers, Raymond Albert
Nature and the crisis of modernity

ISBN: 1-551640-05-5 (bound) –
ISBN: 1-551640-14-7 (pbk.)

1. Human ecology. 2. Biotic communities.
I. Title.

HM206.R64 1994 304.2 C94-9007799-X

Cover Design: Robert Kwak
Book design and layout: Robert Kwak

Mailing Address

BLACK ROSE BOOKS
C.P. 1258
Succ. Place du Parc
Montréal, Québec
H2W 2R3 Canada

BLACK ROSE BOOKS
340 Nagel Drive
Cheektowaga, New York
14225 USA

Printed in Canada
A publication of the Institute of Policy Alternatives of Montréal
(IPAM)

Contents

1 Political Economy and Natural Community 1

Canada's East Coast Fishery: A Case Study in Global Management 4

Outline of Chapters 8

2 Situating the Inquiry 12

Capital 13

Nature 16

The Environmental Crisis and the Crisis of Modernity 18

Modern Social Transformation 19

3 Social Forms of Exchange 24

Archaic Social Forms and Classical Economics 25

Marxist Perspectives on the Evolution of Money 29

Simmel and the Objectification of the Money Form 37

Karl Polanyi and General Purpose Money 43

Psychoanalysis and Money 50

Conclusion 55

4 Cultures of Substitution 62

Money That Breeds 64

The World Eaters 67

Cultures of Substitution 70

Baudrillard and the Political Economy of the Sign 77

Conclusion 85

5 HUMAN IDENTITY AND THE NATURAL WORLD 89

Wild Convergence:
The Sociality of Natural Communities 90

The Residual Cultural Record of
Human-Nature Relations 95

Modern Divergence:
The Split Between Nature and Culture 98

Economic Conditions in
Early Modern England 100

The Outskirts of Meaning:
Nature and Tragedy in *King Lear* 107

The Manufactured World 115

Society as Graveyard 119

Moments of Danger 125

Nature Rendered Unconscious 126

6 PRISONERS OF VALUE:
THE CURRENT ENVIRONMENTAL DEBATE 135

Economic Development as an
Emergent Form 138

Greenback Delusions 144

Accentuating the Gulf 156

7 CONCLUSION: HORIZONS OF SIGNIFICANCE 166

The Aftermath of Collapse 171

BIBLIOGRAPHY 177

INDEX 186

ACKNOWLEDGEMENTS

This book began its life as a doctoral dissertation in the Faculty of Environmental Studies at York University in Toronto, Canada. I would like to thank my advisor Paul Wilkinson, and John Livingston, Neil Evernden, Roger Keil, Ellen Wood, Abraham Rotstein, and Henry Heller for their contribution to the creation of this work. I would also like to thank everyone at Black Rose Books, and especially Robert Kwak for his helpful editorial suggestions. I would also like to thank the Sasakawa Foundation which funds York University's Graduate Fellowship for Academic Distinction for financial support during my time at York. I would also like to acknowledge those who live in Atlantic Canadian coastal communities, and who gave me a context in which to care about the issues addressed here.

But most of all, I would like to thank Laura Jane McLauchlan, without whom I would never have embarked on this book, and our son Lauchlan, who never noticed I was writing it.

1
Political Economy and Natural Community

A great deal of the current discussion related to managing the Earth's resources offers little hope of conserving the natural world because it fails to provide a viable social basis for the relationship between human society and the natural world. This recognition of failure arises out of my experience with the ecologic and social collapse which has occurred in Canada's East Coast fishery, and the similarities that this situation has — both in terms of past history and present goals and strategies — with global management initiatives generally.

I will pursue the argument that environmental issues can be better understood beyond the economic parameters outlined by those who have gathered under the banner of such terms as sustainability. Both the literature on the fisheries and global management repeatedly depict humans as "resources," and nature as a "sink for toxins" or a kind of "factory" which produces an annual surplus for exploitation. Most current conservation initiatives convey only a dim recognition of the extent to which the human relationship with nature has been absorbed by modern economic realities, and do not make problematic a wider set of cultural issues generally referred to as the crisis of modernity. Nor do these representations take into account the literature on international development, especially gender-based analyses, which see Western resource management as a kind of war on local culture and natural habitat resulting from what Gayatri Spivak calls the increasing "financialization of the globe."[1]

I believe that a more viable approach to the current environmental crisis can begin from the contention that the structures and processes of everyday life, and the structures and processes that cause environmental problems, are one and the same. Environmental problems are therefore social and cultural issues which can rarely be separated from their context. As a result, environmental issues cannot be dealt with effectively in isolation. This is especially true of late capitalism wherein the structures and processes of capital and markets have become so dominant that humans only know themselves according to this social context. In order to create a basis for human action that can deal with environmental problems, it is necessary to create a perspective which problematizes the modern structures of everyday life.

The environmental issue with which this book is most concerned is the conservation of the natural world. A social analysis of this issue begins with the assumption that not only is the natural world disappearing in a real sense, but that the human impetus within modern society to conserve what is left of nature, because it is part of "us," is also disappearing. The recognition of this double disappearance — both of nature's complex totality, and the social basis of the relationship between humans and nature — provides a viable starting point for shedding light not only on the conservation issue, but it also locates the environmental crisis in a wider crisis of modernity. This kind of social analysis of the modern perspective would also begin to reflect the misunderstandings characteristic of North-South relationships in the context of international development issues.

To understand the social relationship between nature and culture, I will link the insights of social theorists with those of naturalists who see nature as a social place. For naturalist John Livingston, the sense of "ecologic place" and "belonging" that exists in wild nature is a defining part of this sociality and identity. I believe that Livingston's statement, "in the functioning multispecies community, all participants are subjects,"[2] can form the basis for a social conception of nature that expands the critique of social theorists to include the sociality of the natural world.

When one compares this social conception of nature with Polanyi's argument that in archaic society, human economy is submerged in social relationships based on reciprocity and redistribution, or with what Marx describes as the mutual dependencies of use-values and concrete labour relations, it becomes evident that the insights of such social theorists recognize a wider sociality that links humans and nature. It is this sociality which is not expressed by modern economic realities in which increasingly more participants are made into objects. By linking the insights of naturalists and social theorists, it becomes possible to move beyond the impoverished representations of nature as a sink for toxins or a factory, and to see natural communities in socially viable terms. At the same time, this linkage recognizes nature's social relationship to human community, and allows for the analysis of the commodification of relationships under capital and markets, which are at the heart of current environmental problems.

A glaring example of the social impoverishment of human-nature relations in current conservation discourse comes from the World Commission on Environment and Development's *Our Common Future*::

> They [members of industrial society] draw too heavily, too quickly, on already overdrawn environmental resource accounts to be affordable far into the future without bankrupting those accounts. They may show profits on the balance sheets of our generation, but our children will inherit the losses. We borrow environmental capital from future generations with no prospect of repaying.[3]

This kind of representation of human-nature relations pervades a great deal of conservation discourse and offers little hope of conserving natural communities. The kinds of humans we are, and the kinds of nature we want to conserve, are reflections of each other. If we think of ourselves as human resources available to serve capital, we will only want to conserve natural resources. If we think of ourselves as members of human communities, we can then begin to think about conserving natural communities and habitats. By its very definition, capital is aggressively homeless.

It is the resourcist representation of human-nature relations discussed in *Our Common Future* that is also generally linked with discussions of sustainability, which in turn defines itself as the integration of conservation and development perspectives. This call for integration is a short-circuiting of analysis that takes for granted the processes which undermine social relations. By short-circuiting of analysis, I mean that if conservation is to be integrated with development, then the priorities of development must also be taken into account. This type of a (short-circuiting) approach immediately restores the processes of development and does not make them sufficiently problematic. The vital relationship to understand is the one between the processes of industrial society, and the destruction of embedded relations in human and natural communities. Integrating conservation and development perspectives has little to do with this kind of analysis, and in fact neglects to recognize the transformation in relations which has occurred under capital and markets.

Canada's East Coast Fishery: A Case Study for Global Management

I spent twelve years as a commercial fisherman on Nova Scotia's South Shore. When I began to look into the imminent failure of the fishery and its relationship to wider environmental problems, I was struck by the similarities between the resource management theory and practice used in the fishery, and the current initiatives in global management such as *World Conservation Strategy, Our Common Future, Managing Planet Earth*, and *Caring for the Earth.* I concluded that the goals and strategies set forth to integrate conservation and development in the fishery are identical to those outlined in global management tracts. The complete failure of management frameworks to limit exploitation in the fishery provides a sobering lesson for planners and policy makers who must now devise strategies to avoid the kind of ecological and social catastrophe which has befallen the fishery.

Since selling my boat and licenses in 1991, the sheer enormity of the problems concerning the over-exploitation of marine communities has gradually come to light. Rather than analyse the relationships and processes which have caused the collapse of the fishery (an almost entirely "Made in Canada" problem), most relevant discussions have demonized the foreigner, the seal, and changing climatic conditions which allegedly caused the cod to freeze to death off the coast of Labrador. There has also been much finger pointing by the various sectors within the industry: the inshore blames the offshore, the long-liners blame the draggers, the small companies blame the multinationals, and everyone blames the government. This kind of scapegoating, as well as such technologically contemplated actions as banning the draggers, inhibit a political economy analysis of the relationships between managers and exploiters which would locate the crisis in the fishery within the context of a range of environmental problems. Such actions also fail to problematize some of the general assumptions which inform the relationship between modern society and the natural world.

During the summer of 1993, I took part in the blockade of a Russian ship in Shelburne Harbour to protest foreign overfishing in Canadian waters. At the time, I wrote an article in the local paper entitled, "Should Shelburne County Claim the Fish Stocks?" in which I argued that, like some local cultures in the Southern hemisphere and the aboriginal people of Canada, who reject the dominant Western management regime (in full consciousness of their cultural difference), coastal communities should set out to reclaim their right to regulate their own affairs, especially the right to regulate the fishing grounds adjacent to their communities. Through this kind of argument, I wanted to point out the failure of current management approaches while linking them to global management approaches.

Conserving the natural world is not a process which can be imposed from outside by an external authority at the last minute, as it currently operates in the quota system. In other words, conservation is not an on/off switch for destructive behaviour. Rather, conservation issues have to be in the perspectives of participants who have collectively recognized they are part of both human and natural communities, at all times. This kind of integration at the local level offers some hope of long-term conservation of the fish stocks, although how it will be instituted in the context of an increasingly deregulated global playing field is difficult to say.

In *The Development Dictionary*, Wolfgang Sachs describes resource management approaches, such as those used by the Federal Department of Fisheries and Oceans, as "intellectual ruins" in the Western cultural landscape. For Sachs, most of what would be considered enlightened conservation initiatives are compromised because of their absorption by predatory practices:

> Unwilling to reconsider the logic of competitive productivism which is at the root of the planet's ecological plight, it reduces ecology to a set of managerial strategies aiming at resource efficiency and risk management. It treats as a technical problem what in fact amounts to no less than a civilizational impasse — namely, that the level of productive performance already achieved turns out to be not viable in the North, let alone for the rest of the globe.[4]

Modern culture has to consider the appropriateness of seeing the natural world as a standing reserve from which the huge industrial apparatus can expect a return on its investment. To quote Farley Mowat on the collapse of the East Coast fishery:

> It's a financial vortex. . . . As the thing expands, gets faster, you need bigger and bigger ships, higher technology, and that means greater investment and that means bigger catches and bigger profits. It spirals

> until you reach the point of no return. It's massive high-tech fishing for profit and it has destroyed the resource.[5]

The industrialized expansion of high-technology fishing led to a collapse in the fish stocks off Canada's East Coast during the early 1970s. Up until that time, there was no regulation of this increasingly powerful international fleet of ships beyond a voluntary membership in the International Commission for the Northwest Atlantic Fishery (ICNAF), set up in 1949. The collapse of the stocks, combined with the inability of international agencies to implement regulations in the aftermath of the crisis, led directly to the declaration of the 200-mile Exclusive Economic Zones by coastal states in 1977. As the Canadian government stated in a discussion paper for the Law of the Sea Conference in 1974, "the Canadian government considers customary international law inadequate to protect Canada's interest in the protection of the marine environment and its renewable resources."[6]

It was hoped that nationalizing the fish stocks inside the 200-mile limit would enable coastal states to provide a more stable economic return for domestic fishing fleets, while allowing them to put in place the regulatory infrastructure and biological information base needed to insure long-term conservation of the fish stocks. This was the self-proclaimed mandate of the Canadian government, which had been one of the leaders of the call for a 200-mile coastal zone at the United Nations in the mid-1970s.

To fulfill this mandate, Canada developed its first comprehensive approach to fisheries regulation. Management goals were clearly laid out in *Policy For Canada's Commercial Fisheries*:

> - Institute a coordinated research and administrative capability to control fishery resource use on an ecological basis and in accordance with the best interests (economic and social) of Canadian society.
> - Develop a fully effective capability for the monitoring of information on resource and oceanic conditions, for the surveillance of fleet activity and for the enforcement of management regulations.[7]

Why has this program for economic and biological stability failed so miserably in Canada? The most important thing to understand about Canada's East Coast fishery is that the fish stocks were exploited to the point of collapse before anyone thought that this exploitation should be regulated in any way.

The collapse of the stocks in the early 1970s caused by the expansion of the international distant water fleet, led to the recognition that regulation was indeed necessary. Resource managers tried to gain control over a fully mature industrialized fleet by trying to understand the workings of marine biotic communities that had already been destabilized by over-exploitation. When these factors are combined with the fact that biological information has been gathered for such a short period of time, the vast area being managed, and the inhospitableness of that environment, it becomes understandable why the Federal Department of Fisheries and Oceans has failed in its mandate to provide protection of the fish stocks and economic stability for the industry. It has been a case of the tail trying to wag the dog.

In fact, in terms of Canada's East Coast fishery, resource management has come to mean crisis management. Almost all fishery policy that now exists has come about from inquiries into breakdowns in the industry. Because the policies that came out of these inquiries were implemented in the aftermath of a crisis, they do not reflect attempts to fulfill the twin mandates of biological conservation and economic stability. Rather, they are indicative of conservation policies which are sacrificed to assuage the cries for more fish as the solution to the industry's problems.

The globalisation of the fishing industry that occurred in the late 1960s and early 1970s with the expansion of the international distant water fleets, is now being experienced by many Southern countries in terms of other resources and other sectors of their economies. And once again, it is clear that conserving natural or human communities is not considered a high priority. In fact, the patterns of crisis management and brinkmanship that have occurred in the fishery are being repeated across a wide range of activities, including those of global management.

I believe that the current shutdown in the fishery provides an evocative snapshot of development's "intellectual ruin" (as it has been defined in the context of capital and markets). Some of the significant component parts of this picture are as follows: the assumption that nature exists for human use; the unwillingness to begin initiating conservation measures until imminent depletion of natural communities threaten human prosperity; the assumption that public regulators can and do "referee" private economic activity related to the natural world; the assumption that science can model the interactions of natural communities so as to come up with a magic number which defines sustainable exploitation of natural communities; and the assumption that an external authority can manage human and natural communities so as to promote economic and ecologic stability. The collapse of Canada's East Coast fishery calls these assumptions into question. Furthermore, because these assumptions are central to a wide range of modern human activity, we are returned to the perception that environmental problems are social and cultural issues, which are in turn part of the wider crisis of modernity.

Outline of Chapters

In general terms, Chapter 2 addresses methodological considerations. Chapter 3 and Chapter 4 discuss the emergent processes related to the expansion of capital relations. Chapter 5 presents a residual cultural record of the social relationship between humans and nature which have been undermined by the expansion of capital. Chapter 6 applies this analysis of emergent and residual forms to contemporary discourse on managing the Earth generally, and more specifically to Martin Lewis' *Green Delusions: An Environmentalist Critique of Radical Environmentalism.*[8]

Term definitions and methodological approaches are discussed in Chapter 2 - Situating the Inquiry. Methodology is linked contextually, to questions and problems specific to environmental concerns. Central issues focus on the possibility of representing nature in socially viable terms. This chapter also recognizes the transformation in relations happening in the modern period, and discusses the way such issues influence methodological considerations.

In Chapter 3 - Social Forms of Exchange, I discuss the transformations in social forms of exchange and the ways in which these changes have altered the relationship between human society and the natural world. This is done by examining the work of writers who have discussed the modern realities of political economy from the perspective of changes in economic valuation. This will include the Classical and Neoclassical Schools of monetary theory, Marxian perspectives on commodification and labour, Georg Simmel's presentation of money and objectification, Karl Polanyi's analysis of general purpose money, and Norman O. Brown's psychoanalytic perspective on economics and history. This chapter will provide a context for what will be discussed in chapter 4 with regard to contemporary social realities, which Jean Baudrillard has referred to as the political economy of the sign. The essentially economic process of valuation (as discussed in chapter 3) becomes, in the political economy of the sign, an inclusive kind of reality where not only money, but all aspects of contemporary society become part of a culture of substitution.

The main focus of Chapter 4 - Cultures of Substitution is on Jean-Joseph Goux's discussion of cultures of substitution and Jean Baudrillard's analysis of the political economy of the sign. The disruption of forms of social relations occurring alongside the expansion of capital, becomes universalized in the political economy of the sign. It is this generalized disembeddedness which encloses modern human society in a circular self-referentiality, and erases a social basis for human-nature relations.

Norman O. Brown describes history as the triumph of dead matter over the living. The transformation of use-value into commodity, of commodity into money, of money into a general equivalent, of the general equivalent into generative capital, and of generative capital into a separation of value and sign from the things they represent, have charted an archaeology of the economic value form which confirms Brown's statement. Money's ultimate triumph is that it ceases to be specifically economic at all. Instead, it is entirely based on expressions of value. I contend that it is this value form that is promoting the destruction of the natural world, and also inhibiting (through its enigmas and mystifications) the kind of analysis which would identify it as doing so.

Chapter 5 - Human Identity and the Natural World argues that the relationship between humans and nature is a social one, and that there is a residual cultural record of the ways in which this relationship has been undermined by the processes of capital in the modern period. In order to illustrate this social transformation, I identify three moments of danger in the modern period where there is a particularly intense residual expression of loss. These moments of danger are: Early Modern Europe and the loss expressed in Tragic literature; the Industrial Revolution and the dislocation recorded in Romantic poetry; the globalisation of capitalism and radical environmentalism's rejection of that transformation.

Only when the natural world is seen as a social place does it become possible to link the crisis in human identity to the environmental crisis as part of a more widespread crisis of modernity. It is this kind of approach which connects radical environmentalism with other expressions of loss of sociality in the modern period. It is also this kind of approach which recontextualizes the process referred to as the social construction of reality as its opposite; that is, the gradual undermining of a sociality which has its legacy in the natural world.

Chapter 6 - Prisoners of Value: The Current Environmental Debate applies the analysis of modernity set forth in Chapters 3, 4, and 5 to current discourse on managing the Earth. More specifically, I discuss Martin Lewis' book and the way this work records the moment of danger related to the globalisation of capitalism. While Lewis' book has much in common with other works of moderate environmentalism, it is particularly apt for the discussion here. Lewis sets out to refute the arguments of radical environmentalism, and to show that its assumptions about the social basis upon which human society's relationship to nature is built, are mistaken.

I argue that Lewis' promotion of the idea of "decoupling" human society from nature in order to save nature, is a recognition that human identity no longer needs nature to define itself. I contend that decoupling is the completion of the disembedding processes which I have argued are at the centre of both the environmental crisis and the crisis of modernity. What I have presented in terms of loss, Lewis sets forth as an ascendant solution.

In the Conclusion - Horizons of Significance, I distinguish between the instrumental and strategic way that Lewis uses "decouple," and the emotive sense of loss which the word "decouple" engenders in those who recognize what this transformation represents. Charles Taylor and Livingston present the demise of social claims which the world can make on humans, as exemplary of the increasing impoverishment of modern social forms. I relate this analysis to the plight of Atlantic Canadian coastal communities.

NOTES

1. Gayatri Spivak. February 1993. "Can Discourse Be Countered?" lecture given at York University.
2. Jonathon Livingston. 1994. *Rogue Primate: An Exploration of Human Domestication*. Toronto: Key Porter Books, p. 111.
3. World Commission on Environment and Development. 1987. *Our Common Future*. New York: Oxford University Press, p. 8.
4. Wolfgang Sachs. 1992. *The Development Dictionary*. London: Zed Books, pp. 35-6.
5. Farley Mowat. 1990. Quoted in "Hibernia Blues" by Glen Wheeler, *Now Magazine*, Vol. 10, No. 4, p. 10.
6. Canada. 1974. *Law of the Sea Discussion Paper*. Ottawa: Department of External Affairs, p. 3.
7. Fisheries and Marine Service. 1976. *Policy for Canada's Commercial Fisheries*. Ottawa: Department of the Environment, p. 21.
8. Martin Lewis. 1992a. *Green Delusions: A Environmentalist Critique of Radical Environmentalism*. Durham: Duke University Press.

2
Situating the Inquiry

> . . . what counterstrategy is available for the illumination of reality that does not in some subtle way replicate its ruling ideas, its dominant passions, and its enchantment of itself? As I see it, this question is both necessary and utopian. . . . For this reason, what I call negative criticism is all that is possible, apt, and demanded at an intellectual level.
>
> *Michael Taussig*[1]

Modern Western society has become a de-animating culture. It is a world in which living things die to make a dead thing grow. To characterize the achievements of modernity as the growth of dead things is to risk a lot of wrath. To characterize the achievements of modernity as the growth of dead things is to also implicate myself in that destruction because capital is not a thing, but a set of processes and relationships which, through its necessity to expand, draws life out of humans and nature. But, these processes and relationships have been created and maintained by humans. Furthermore, a self-conscious recognition of that generative role is essential for modern humans to understand and respond to environmental problems in a way that is not captive to these de-animating processes.

This inquiry is not about appropriate functional knowledge which is utilized in the name of problem solving and management. This inquiry is about dysfunctionality and dislocation and loss. In broad terms then, this work is a record of animus leaving the world. Central to this leaving is the role of ideas, values, and theories which have recorded the displaced trajectory of that departure. Anyone who has seen or felt a baby being born, or had a whale come up under a boat, understands that ideas are no big deal. Ideas come and go, like rain in the ditch at the side of the road. It is a defining aspect of modernity to treat ideas and values as matters of utmost importance. Within that importance however, the immanence of nature has lost its significance. The enlivening of a deadening process that grows, has been the eugenics project of modernity.

Whatever it has meant for other groups with regard to the natural world, modernity has meant only the intensification of exploitation. What has not been swept aside as an irrelevancy has been homogenized into ever-expanding monocultures which serve capital and markets. I recognize that this is a particularly dark view of the modern period. If I am presenting an especially negative representation of life under capital and markets, it is because there is an element of the corrective here. Most discussions about the conservation of nature take no account of the implicitly predatory aspect of modern relationships. It is my purpose to highlight the destructiveness of these relationships. The starkness of my representation forces a recognition which creates room for discussing alternate forms of relationships. That said, I think it is all but impossible to exaggerate the seriousness of a great many environmental problems, or to underestimate how little has been done in the name of dealing with these problems.

The scope of the material covered in this book is wide, both in terms of the time period covered and the range of perspectives discussed. It may seem at times that I am oblivious to the usual disciplinary barriers that exist in a great deal of work which comes out of universities. The works that I have chosen to discuss here — be it Marx, Shakespeare, or Baudrillard — are selective and symbolic of what I consider to be important aspects of modernity. I am aware that at every turn, there are ongoing controversies and differing points of view around each of the works I discuss. For the most part, I have not referred to these discussions, for to do so would have diverted the focus of the book away from the arguments I am trying to make. This book is first and foremost about the conservation of the natural world, and modernity is analysed with this focus in mind. This is not defensive minded work but rather, an attempt to articulate a version of modernity that sheds light on the troubled relationship between modern humans and the rest of nature.

Capital

> Not a single line of these investigations is meant to be a statement about economics. That is to say, the phenomena of valuation and purchase, of exchange and the means of exchange, of the forms of production and the values of possession, which economics views from *one* standpoint, are here viewed from another. . .

> money not merely reveals the indifference of purely economic techniques, but rather is, as it were, indifference itself, in that its entire significance does not lie in itself but rather in its transformation into other values.
>
> *Georg Simmel*[2]

> It is always through replacement that value is created.
>
> *Jean-Joseph Goux*[3]

The seeming intractability of the juggernaut of human exploitation of nonhuman nature, is not merely a problem of harnessing greed and profit. It is as much motivated by complicity and resignation. In fact, what is most haunting about the rapacity of the human exploitation of nature is not the hot-blooded greed of destruction, but the indifference with which it is viewed by industrial society. In *The Arrogance of Humanism,*[4] David Ehrenfeld refers to this as a kind of double bind for nature: either nature is identified as a resource and hence exploited, or it is swept aside as being irrelevant.

Jean-Joseph Goux outlines this kind of archaeology of value where value forms which originally were found within commercial transactions, now appear in unexpected ways in a wide range of human activity:

> My first analyses, then, arose from the theoretical conviction that a certain style of thought could establish connections among semiotics, psychoanalysis, and economics. . . . It then appeared to me, more specifically, that these connections could be conceived in terms of the phenomena of exchange, for the semiotic, economic, and psychoanalytic all emphasized the question of substitution and its correlative, value . . . . Thus the social sciences (from economics to anthropology) as well as their philosophical impact were emphatically implied by the notion of exchange . . . . This operation led to the nuclear and, as it were, metasocial aspect of the symbolic function, to what can be designated from different angles as the "in place of," "stand-in," or "supplement," the constant that set apart, equivalence posited despite difference,

> the maintenance of an ideality throughout changes in materiality, etc.[5]

Although he does not discuss the destruction of the natural world specifically, I believe that Goux's conclusions about value have profound ramifications for understanding the relationship between modern humans and nature. Goux's approach also provides a basis for expanding an analysis of economics beyond the more narrow considerations of production and profit, to more wide-ranging questions concerning the relationship between quantitative and qualitative values.

I contend that there can be nothing more destructive to the perceived significance of human relationships with the natural world than the pervasive cultural assumption that objects gain value based on their exchangeability with other objects according to an abstracted unit of analysis or measurement. If it is true that (as in the case of capital) this measurement is a dynamic that requires those units of measurement to expand continually through the conversion of more and more material goods into commodities for exchange, then it becomes possible to begin to understand what it is that makes environmental problems seem so intractable. Capital is an abstracting and extracting process which only runs the gauntlet of materiality so that it can return as more abstraction. This process is at the heart of modern culture and is also at the heart of environmental problems.

When I make these claims about capitalism, I do not mean to say that human-nature relations would be necessarily unproblematic without capitalism, or that alternate cultures are somehow more humane. In fact, the transformation from Paleolithic culture to Neolithic culture, the beginnings of the domestication of plants and animals, was at least as significant a change in relations, as the shift to capital and markets. An analysis of the transformation in conceptions of deities in early cultures, from dispersed immanence throughout nature to abstracted and universalized monoliths, reflects the hardening of the self-other dichotomy which is at the heart of the processes by which nature is objectified.

I have chosen the modern period because it lends itself to a coherent basis for analysis. Also, contemporary conservation discussions reflect these modern assumptions and so focusing on them is directly relevant to my purposes. To focus on the destructive and predatory aspects of modernity is not to deny that there have been many positive developments in modernity as well. In every life which is lived, there is always the possibility of affirmation and coherence. In the modern period, this affirmative possibility has been expressed, more often than not, in spite of, or in resistance to, the more destructive aspects of the expansion of capital and markets.

Nature

> I no longer believe that there is, in practice, such a thing as a "renewable" resource. Once a thing is perceived as having some utility — any utility — and is thus perceived as a "resource," its depletion is only a matter of time.
>
> *John Livingston*[6]

> We are regaled daily with the promises of fresh solutions to the "environmental crisis"; yet it is difficult to suppress a sense of déjà vu. . . . Indeed, many environmental tracts published early in the twentieth century could very nearly be used in contemporary classrooms. One cannot avoid the sense that however much our environmental awareness has increased and our intentions to "save the Earth" improved, at root nothing has changed.
>
> *Neil Evernden*[7]

My analysis rests on the assumption that nature is a social place. Nature is not a resource for human monetary wealth. Nature is not a large farm. Nature is not an ecosystem — a term which I regard as the conceptual partner to economic globalisation. Nature does not have goals ("survival of the fittest"), nor can it be said to be driven by any one thing (like a selfish gene). Nature is local, manifold, participatory, and cyclical. In other words, a community.

The separation of nature and culture is an acknowledgment that certain forms of sociality which humans and nature once shared, have now disappeared from human society. Because human society has become increasingly homogenized by economic considerations, it also reconceptualizes nature as now having lost complexity. Hence the origin of the double disappearance. This split intensifies throughout the modern period until we arrive at the arid dichotomy of humans and natural resources, or should I say, capital and resources (both human and natural). In this present context, there is not much in the relationship between humans and the natural world that can be described as social.

In *Nature and Madness*, Paul Shepard states that "the real bitterness of modern social relationships has its roots in the vacuum where a beautiful and awesome otherness should have been encountered."[8] Social theorists such as Karl Polanyi and Jean Baudrillard have attempted to describe this loss of identification which has occurred as a result of changes brought on by the strengthening of modern economic realities. Only by denying nature a social standing can humans presume to socially construct meaning on their own.

The qualities of natural sociality — and this could be regarded as my definition of nature throughout this work — are participatory and communal. Livingston describes a natural community in this way:

> At the level of the community, 'other' is bereft of. . . abstract meaning or utility. . . . In the functioning multispecies community, all participants are subjects . . . . there need be no other; the community is a whole unto itself.[9]

To say that all participants are subjects and that no group of participants is privileged or another objectified, is to begin conceiving of a viable social representation of nature. I contend there is a connection between this kind of sociality and the sociality which theorists like Polanyi and Baudrillard point to in their discussions about such terms as reciprocity. It is this definition which is also in sharp contrast to the current resourcist perspective in which all participants are objects.

In his article "The Incarceration of Wildness: Wilderness Areas as Prisons," Thomas Birch describes wildness as a "nonconflictive, complementary relationship with otherness" in which a "radical openness" makes identities "contingent" and "unfinalized."[10] Birch contrasts this understanding of wildness with "the central myth of Western culture" where "civilization brings light and order to the wild darkness of savagery — the legitimizing story that cloaks conquest, colonization, and domination." In my discussion about the transformations of the money form within modernity, I will investigate one of the central myths of Western culture related to the expansion of capital. I do so in an attempt to understand why conservation of nature continues to lose ground.

I recognize the difficulty inherent in criticizing certain depictions of nature as representative of particular social realities, while presenting an alternate depiction of nature which is not somehow subject to the same ideological biases. In the struggle to find ways of talking about what is wrong with current human-nature relations, there is a tendency to present alternate views of nature which are "better." Such tendencies inevitably become vulnerable to being accused of creating a Romantic vision of nature from which humans have "fallen." I have attempted to avoid this "virgin-birth" depiction by treating nature as a social place amenable to the insights of social theorists. Nature in effect becomes the social, and regardless of whatever else nature may be, there is still a lot that social theory can say about objectified and exploitive relationships. In pursuing this line of argument, I am aware that (as Taussig states in the opening quotation to this chapter) the only possible counterstrategy against enthrallment is a kind of negative criticism characterized by the continual "make and break" inherent in disarming methodologies.

The Environmental Crisis and the Crisis of Modernity

> Machine production in a commercial society involves . . . a transformation. . . of the natural and human substance of society into commodities. The conclusion, though weird is inevitable. . . . [T]he dislocation caused by such devices must disjoint man's relationships and threaten his natural habitat with annihilation.
>
> *Karl Polanyi*[11]

I will argue that unstable conceptions of human identity — as expressed in discussions of "the death of the subject" in literature about the crisis of modernity — and the failure of initiatives to conserve nature as expressed in radical environmentalism, are related to the way pervasive economic categories inform modern understanding of both humans and nature.

This environmental crisis then, is a social crisis related to the viability of identity resulting from the destruction of habitat and the increasing pervasiveness of monocultures. I do not approach crisis in a problem-solution manner which would require instrumental tinkering with carrying capacity and toxin levels. I approach crisis as a social event which is perceived by those who claim to experience changes in social reality as a crisis. I therefore link the environmental crisis to the crisis of modernity by describing them as expressions of a collapse in social relationships and social identity.

Although this book will not attempt to take into account the voluminous literature on the crisis of modernity, it is nonetheless meant to be (in its analysis of human-nature relations) a comment on, and a direct contribution to, that discussion. Specifically, my analysis contends that the pervasive provisionalness of the social construction of meaning which recurs in the body of literature about the crisis of modernity, represents a massive *social* failure of relationships which is historically specific to late capitalism. This social failure is endemic to modern Western contemporary culture and is not universal to all cultures at all times.

Modern Social Transformation

The goal of this book is to analyse current discourse on managing the Earth. This analysis is carried out by examining the interrelationships between the social basis of human identity, and the corpus of social representations of the natural world. One of the most significant ramifications of this kind of social approach is the dissolution of the standard separation between nature and culture. While culture is usually viewed as a kind of *Realpolitik* where activity and interaction are scrutinized in a self-conscious way by the social sciences, nature is seen as the more predictive and reductive interplay observed by biologists and ecologists, as is expressed in the usual nature-nurture debates about identity. If there is a separation between nature and culture in my discussion, it is because there is an increasing homogenization and standardization of human culture as it is mediated by the forces of technology and commodification. The end result is that nature becomes objectified by these forces. What is interesting about the separation between nature and culture is that the reduced standing of nature mirrors human culture, but without the dross of mystification which usually hides this intensification of exploitation from human participants.

The social basis of my analysis provides a critique of capital. More specifically however, it allows a critique of those discourses on managing the Earth which take capitalism for granted. Broadly speaking, this social critique of capital makes use of Polanyi's discussion about the historical transformation of social relations, from embedded to disembedded relations:

Generalized Embeddedness

Anthropologists such as Marshal Sahlins characterize economic relations in early and archaic human societies as embedded in social institutions. That is, the activities associated with the gathering and distributing of food were in every way social activities and hence, cannot be described as autonomously economic in the modern sense. It is this understanding of generalized embeddedness that allows us to expand discussion to include a consideration of the natural world as something that is socially embedded. But in fact, the context of generalized embeddedness precludes a discussion about nature as an entity separate from culture because ultimately, this division does not exist.

The Dislocating Trajectory of Modernity

Although there have been transformations in social relationships dating back to the Neolithic period, the accepted representations of the processes of modernity intensified the trend toward separation and division: church from state, economics from religion, the individual from the corporate body, god from the Earth, rationality from superstition, the sign from the signified, nature from culture. The expansion of the relations associated with capital are central to this process. In environmental terms, modernity represents first the appearance of nature separate from culture, followed by the gradual withdrawal of social significance from the natural world.

Generalized Disembeddedness

The various processes of separation associated with modernity move from the economic sphere to become generalized and universalized across a wide range of activities. Such theories as

Baudrillard's political economy of the sign and Goux's cultures of substitution, argue that this separation or disembeddedness of valuation from "the things themselves," is so pervasive that analysis becomes enclosed within a chain of signifiers and never refers back to what I would contend would be an embedded set of social relations. By using the terms "the things themselves," I do not mean that there is an authentic and definitive essence which was rendered invisible by an increased attention to signs, but rather that a particular set of relations loses expression.

If the dislocating trajectory of modernity is defined in terms of separation, generalized embeddedness is defined by the collapse of separation into the self-referentiality of an abstract code which no longer refers back to the relationships between the things themselves. This is the social failure which is expressed in the literature on the crisis of modernity.

Methodological approaches to this transformation of social relations can be described in terms of Raymond Williams, Michael Taussig, and Walter Benjamin's works. Williams presents a view of social change which attempts to move away from reductive base-superstructure arguments — which would claim that economic relations define culture — to one that sees emergent and residual forms of ideas, practices, social relationships moving into and out of a dominant paradigm. For Williams, these hegemonic structures are not monolithic. Instead, they "are highly complex, and have to be continually renewed, recreated and defended" against the emergent and residual forms which contradict the dominant one.[12] By emergent forms, Williams means that "new meanings, new practices and experiences, are continually being created." In contrast to emergent forms, residual forms are "experiences, meanings and values, which cannot be verified or cannot be expressed in terms of the dominant culture" because they are the residue of a "previous social formation." I will argue that the expansion of capital is an emergent and defining form of modernity, while other forms of social relations which are subverted by capital, are residual forms lost to the dominant paradigm.

In Michael Taussig's discussion of tin miners and plantation workers in South America, there is an emphasis on the cultural losses incurred as a consequence of modern capital. This can be related to Williams' residual forms. Taussig states:

> Let us explore this notion that the[re] are collective representations of a way of life losing its life, that

> the[re] are intricate manifestations that are permeated with historical meaning and that register in the symbols of that history. . . . It would be a shocking oversight not to realize that these beliefs occur in a historical context in which one mode of production and life is being supplanted by another.[13]

Taussig's argument that there is a collective representation of "a way of life losing its life," is central to my social analysis of capital. I contend that there exists a residual cultural record of forms of social relations being lost as a result of the emergence and expansion of forms of relations pertaining to capital. In residual terms, modern history is a record of appearances which are temporarily discussible as they are lost. Social forms, and the values and relationships associated with those forms, move from an embedded invisibility, to temporary discussibility (fracture and dislocation), and finally, to disembedded invisibility, at which point they are no longer recognizable because the relations on which they depend have changed.

The context of fracture and dislocation can be expressed with reference to Walter Benjamin's "Theses on the Philosophy of History," where he states that "to articulate the past. . . means to seize hold of a memory as it flashes up at a moment of danger."[14] Building on this conception of social transformation within modernity, I contend that expressions of emergent and residual forms gain increased clarity and intensification in particular moments of danger when the processes of change become more rapid and dislocating.

To contrast emergent and residual forms is not to say that there are only two doors leading onto and off the stage of history, and that contending characters definitively represent one form or the other as they struggle for that stage. The historical process is too root-bound and hybridized for that kind of clarity. But this representation nonetheless is supported by the fact that there has been opposition and resistance in every heralded transformation of modernity. It is from this residual cultural record of resistance and refusal that it becomes possible to make visible the embedded social relationships which have linked human communities to natural communities.

NOTES

1. Michael Taussig. 1980. *The Devil and Commodity Fetishism in South America*. Chapel Hill: University of North Carolina Press, p. 7.
2. Georg Simmel. 1990. *The Philosophy of Money*. New York: Routledge, p. 54.
3. Jean-Joseph Goux. 1990. *Symbolic Economies*. Ithaca: Cornell University Press, p. 9.
4. David Ehrenfeld. 1978. *The Arrogance of Humanism*. New York: Oxford University Press.
5. Goux (1990:2).
6. John Livingston. 1981. *The Fallacy of Wildlife Conservation*. Toronto: McClelland and Stewart, p. 43.
7. Neil Evernden. 1992. *The Social Creation of Nature*. Baltimore: Johns Hopkins University Press, p. ix-x.
8. Paul Shepard. 1982. *Nature and Madness*. San Francisco: Sierra Club, p. 108.
9. John Livingston. 1994. *Rogue Primate: An Exploration of Human Domestication*. Toronto: Key Porter Books, p. 111.
10. Thomas Birch. 1990. "The Incarceration of Wildness: Wilderness Areas as Prisons." *Environmental Ethics*. Spring Vol. 12, No.1, p. 9.
11. Karl Polanyi. 1957. *The Great Transformation*. Boston: Beacon Press, p. 42.
12. Raymond Williams. 1980. *Problems in Materialism and Culture*. New York: Verso, p. 38.
13. Taussig (1980:17-18).
14. Walter Benjamin. 1973. *Illuminations*. London: Collins, p. 257.

3
Social Forms of Exchange

> Cultural change is not understood as unfolding according to some predetermined logic (of development, modernization, or capitalism) but as the disrupted, contradictory, and differential outcomes which involve changes in identity, relations of struggle and dependence, including the experience of reality itself.
>
> *Aihwa Ong*[1]

> Money represents the divine existence of commodities.
>
> *Jean-Joseph Goux*[2]

This chapter's goal is to offer an analysis of the appearance of various forms of exchange that preceded the social forms of capital. It is also my goal to do so without discussing these appearances in terms of developmental stages that might grant a sense of inevitability to them. This sense of inevitability is conveyed by such terms as "developing countries," and in the general triumphalism that now sees market economy as the end of history. The twisted logic of this approach to the natural world generates such terms as "underutilized species" in the fishery. It also limits a great deal of the conservation debate between the North and South by assuming that modern economic frameworks represent the goals of the rest of the world.

To confer inevitability on these modern social transformations is also to suggest the primacy of the present. A recognition of the seriousness of current environmental problems highlights the inappropriateness of much of what is occurring in modern industrial society. This inappropriateness does not confer necessity on the present, but instead leads to an analysis of modern realities as the root causes of social transformations which have led to a wide range of environmental problems.

I will present these transformations of the social forms of exchange and the ways in which they have altered the relationship between human society and the natural world, by examining the work of writers who have discussed the modern realities of political economy from the perspective of changes in economic valuation. This will include the Classical and Neoclassical Schools of monetary theory, Marxian perspectives on commodification and labour, Georg Simmel's presentation of money and objectification, Karl Polanyi's analysis of general purpose money, and Norman O. Brown's psychoanalytic perspective on economics and history.

To a great extent then, this chapter will provide the background for what will be discussed in the next chapter with regard to contemporary social realities which Jean Baudrillard has referred to as "the political economy of the sign." What in this chapter is a description of an essentially economic transformation of valuation becomes in the political economy of the sign, an inclusive kind of reality where not only money, but also most aspects of contemporary society, become parts of a culture of substitution.

Archaic Social Forms and Classical Economics

Typical of the universalizing tendency of the Neoclassical approach to modern capital is Thomas Guggenheim's comment in the midst of a discussion about Ancient and Chinese philosophers: "It seems that a relationship between the quantity of money available and the general level of prices was perceived quite early in the history of economic thought."[3] The analysis which underpins his argument begins and ends with a movement toward "the advantages of the monetary tool." This privileging of certain dynamics in history is the basis for explaining a universal developmental process which finally leads to a concept of market economy.

Within discussions about the Classical perspectives on the evolution of money, there is a general tendency to privilege present economic realities so that the history of economic thought appears to be an inevitable movement toward the present. This tendency in effect affirms Adam Smith's oft-quoted phrase concerning "the general disposition to truck, barter, and exchange" leading to the increased convenience and efficiency of money.[4] In his chapter in the *Wealth of Nations* entitled "Of the Origin and Use of Money," Smith describes a set of social conditions and processes which inform commercial activity in a way that is not much different from Marx's ideas discussed later in this chapter:

> When the division of labour has been once thoroughly established, it is but a very small part of a man's wants which the produce of his own labour can supply. He supplies the far greater part of them by exchanging that surplus part of the produce of his own labour, which is over and above his own consumption, for such parts of the produce of other men's labor as he has occasion for. Every man thus lives by exchanging, or becomes in some measure a merchant, and the society itself grows to be what is properly a commercial society.[5]

Where Smith and Marx differ is that while Smith sees every man as "some measure a merchant," Marx's formulations involve class divisions and the alienating aspects of abstract labour and exchange-value. Smith goes on to state that in this commercial society, because of the division of labour, each person "has more of a certain commodity than he himself has occasion for, while another has less." To overcome the "inconvenience" of a situation where "they have nothing to offer in exchange," each person endeavours to have "a certain quantity of some one commodity or other, such as he imagined few people would be likely to refuse in exchange for the produce of their industry." Although many commodities have served this purpose of exchange throughout history, "men seem at last to have determined by irresistible reasons to give the preference, for this employment, to metals above every other commodity." Hence the widespread use of precious metals as a commodity which is synonymous with money.

For Smith, these "irresistible reasons" mesh with the logic of increasingly complex economic arrangements in which more developed forms of money play an indispensable role. From this point, all discussion of the money form as represented by the Classical school, is in functional terms related to how best to maximize the efficiency of economic activities where money plays a central role. Likewise, in *The Theory of Money*, Jurg Niehans states that "the most fundamental questions of monetary theory concerns the services or 'functions' of money."[6] The most significant of these services is to initially free society from "the shackles" of the barter system so that there can be an "increase in welfare" based on the increased ease of economic activity.

As Crump states in *The Phenomenon of Money*, "by taking the institutions for granted, the monetary theorist is seduced into accepting, as axiomatic, a number of statements about money, which are at most true only in a limited range of monetary systems."[7] For example, this functional perspective provides little by way of a basis from which to analyse the current dysfunctional aspects of social relations between modern humans and the natural world.

A quotation which illustrates this process of universalizing present conditions back into archaic and tribal societies comes from de Coppet's description of the relation between money and the natural world in the Solomon Islands:

> The ritual [of exchange] prolongs their [Solomon Islanders] effects beyond the moment of time, to integrate them in both the past and the future. The circulation of these moneys is subject to precise rules, so that, together with men, women, and children and other goods recognized in the local culture, they form a system of exchanges which maintains and perpetuates the established patterns of social organization. The implicit immortality of the society, as such, is thus maintained by the mortality of the people and goods which momentarily cross its path. Both the living and the dead combine in the eventual destruction of all things, so that in the end nothing remains save these strings of money, and the unceasing ballet which they perform. These moneys, the tangible supporters of the law, are all that remain of the ancestors, and as such they are the all-powerful accomplices to the process of time.[8]

Described in this way, money seems to inevitably lead to the destruction of the natural world. The immortal bulwark that money provides against the processes of time, works in opposition to, and devalues, the ever changing and ever renewing cycles of the living world. In fact, de Coppet is reading the process of abstract valuation back into an archaic structure without due regard for the legitimacy or relevance of such an analysis.

The attachment to "useless" objects such as money in traditional societies is qualitatively different from the valuation which operates within capitalism. This was made clear by Malinowski's famous article on the *Kula* trade of shells among the Trobriand Islanders: ". . .the temporary ownership allows him to draw a great deal of renown, to exhibit his article, to tell how he obtained it and to plan to whom he is going to give it."[9] Taussig contrasts this fetishism in pre-capitalist societies with that of market-based societies:

> [T]he fetishism that is found in the economies of pre-capitalist societies arises from a sense of organic unity between persons and their products, and this stands in sharp contrast to the fetishism of commodities in capitalist societies, which results from the split between persons and the things that they produce and exchange. The result of this split is the subordination of men to the things they produce, which appear to be independent and self-empowered.[10]

In other words, the relation to objects in pre-capitalist societies is integral to social relationships, whereas the human relation to objects in capitalist societies tends to be alienating, and ultimately undermines social relationships. The social relationships of commodities is generalized across society and therefore negates the particular history necessary for a social relationship between humans and things.

Because monetary realities seem to operate in an autonomous and self-defining fashion in industrial society, this autonomy is often read back into earlier forms of social relations. Whereas modern economic realities are understood generally in term of their autonomy from, and sometimes antipathy toward, social and cultural institutions, the economies of archaic societies were deeply rooted in social and religious institutions. To quote Polanyi:

> Man's economy, as a rule, is submerged in his social relationships. He does not act so as to safeguard his individual interests in the possession of material goods; he acts so as to safeguard his social standing, his social claims, his social assets. He values material goods only in so far as they serve this end.[11]

The rejection of the application of modern economic categories to archaic societies is also conveyed by Marshal Sahlins in *Stone Age Economics*:

> The primitive order is generalized. A clear differentiation of spheres into social and economic does not there appear. . . . [E]very exchange as it embodies some coefficient of sociability, cannot be understood in its material terms apart from its social terms.[12]

Becoming sensitive to the inappropriateness of reading modern economic rationale back into other historical periods, is central to discovering how to talk about the ways in which present conditions have caused environmental problems. Analytical space collapses if current conditions are seen as natural outcomes of human propensities or an expression of the order of things.

Marxist Perspectives on the Evolution of Money

The contribution that Marx can make to an understanding of the destruction of the natural world by industrial society, involves the three-fold linkage of human alienation, a resourcist view of nature, and the emergence of the abstract-value form. In the opening section of *Capital,* Marx describes how these processes occur in the context of commodity production by tracing the evolution of the money form.

Capital begins with the statement that since the measurement of wealth under the capitalist mode of production is the accumulation of commodities, "our investigation must therefore begin with the analysis of a commodity" since "the simple commodity form [where objects are exchanged] is the germ of the money-form." Marx then goes on to state that a commodity is "an object outside us, a thing that by its very properties satisfies human wants of some sort or another." Within the cosmology of commodities, the natural world is swallowed up as a "useful" aspect defined by human priorities: "A commodity, such as iron, corn, or a diamond, is therefore, so far as it is a material thing, a use-value, something useful."

The commodification of the relationship between "us" and what is "outside us" is of particular interest to a perspective that is concerned with the destruction of the natural world by industrial society. This commodification process is also central to the invisibility of nature and the mystification that shrouds modern social realities and their destruction of natural communities. To reveal this invisibility and mystification, Marx engages in a genealogy of the value form which reveals "the enigma of money" representative of the value form's ascension to power, and the institutionalization of its role in human society.

Marx engages with the "enigma of money" by presenting the evolving social relationships between commodities. The use of the phrase "social relationships between commodities" cannot be over-emphasized when discussing the realities of modern industrial societies. It is the commodity which is becoming increasingly more significant in forging social relationships between humans and nature. Hence, Marx can state that ". . . value can only manifest itself in the social relation of commodity to commodity."

In general terms, the commodity genesis of money begins with the elementary relationship between two commodities (two pigs for a cow), followed by more complex relations between numerous commodities (a small town market), which leads to a few select commodities becoming privileged in the interplay of these transactions (the selection of a particular commodity that may serve as local money). In the final stages, there is the monopoly of an exceptional commodity such as gold or silver in which all value seems to find expression. Marx sums this process up in this way:

> The primary or isolated relative form of value of one commodity converts some other commodity into an isolated equivalent. The expanded form of relative value, which is the expression of the value of one commodity in terms of all other commodities, endows those other commodities with the character of particular equivalents differing in kind. And lastly, a particular kind of commodity acquires the character of universal equivalent, because all other commodities make it the material in which they uniformly express their value.[13]

Within the original elementary form of value in which "the whole mystery of the form of value lies hidden," Marx outlines the two poles of relative form and equivalent form where the value of one commodity (relative form) is expressed in terms of another commodity (equivalent form) as in the statement, twenty yards of linen equals one coat:

> The relative form and the equivalent form are two intimately connected, mutually dependent and inseparable elements of the expression of value; but at the same time, are mutually exclusive antagonistic extremes — i.e., poles of the same expression. . . . The value of the linen can therefore be expressed only relatively — i.e. in some other commodity. The relative form of the value of the linen pre-supposes, therefore, the presence of some other commodity — here the coat — under the form of the equivalent.[14]

Presenting this evolution of the money form as part of a developmental process, Marx sees this stage as an inefficient and incomplete "germ," which is metamorphosed by forces of commodity relations. If there is any developmental process in history, it relates to the increasing demands of commodity relations caused by this metamorphosis, and not to any innate propensities of human identity such as the desire to compete or to maximize efficiency.

The second stage of total or extended value form in the relationships of commodities expands what had been a particular relationship between two commodities [linen for a coat] in the elementary or accidental form, into a more generalized reality in which a wide range of commodities are brought into market relations with each other. As Marx states, this form of commodity ". . . by virtue of the form of its value, now stands in a social relation, no longer with only one other kind of commodity, but with the whole world of commodities. As a commodity, it is a citizen of the world." Once again, Marx emphasizes the social nature of commodities by referring to them as "citizens."

In the third stage of the social relations of commodities, the various relationships between commodities are now expressed in terms of one commodity only. The values of commodities arrive at a standardized, "socially recognized form" as they enter into universal circulation. Goux writes this about the stage in which one commodity becomes a universal standard:

> The choice [the uninominal election] of a general equivalent thus gives the world of commodities an essential change of face. By exchanging their intricate mutual dependencies for a simple relationship to a single equivalent, by expressing their value in a single mirror, they acquire a determination that is both social and autonomous.[15]

In this transformation from "intricate mutual dependencies" to the single mirror of a general equivalent, the monolithic character of modern economic realities becomes increasingly more defined according to the sociality of commodities which are equated in terms of a single measure. The equivalent form — in whatever commodity it is embodied — assumes a privileged position and seems "endowed with the form of value by Nature itself." This role of being the measure of other commodities creates "the enigmatic character of the equivalent form," which has been associated with commodities such as gold and silver. In this role, gold and silver domesticate the world by converting it into standard measurements.

The metamorphosis of gold as a singled out commodity, to gold as a universal equivalent defined in terms of money, represents the fourth stage in the evolution of the money form. In its incarnation as money, gold is "expelled into transcendence" because it is a monopolizer of the value form. To quote Marx:

> . . . if a commodity be found to have assumed the universal equivalent form, this is only because and in so far as it has been excluded from the rest of all other commodities as their equivalent . . . . The particular commodity, with whose bodily form the equivalent form is thus socially identified, now becomes the money-commodity, or serves as money. It becomes the special social function of that commodity, and consequently its social monopoly, to play within the world of commodities the part of the universal equivalent.[16]

The importance of such transformative processes cannot be overestimated when discussing the relationship between human society and the natural world. The expelling of a standardizing power of measurement promotes homogeneity of relations and inhibits particular social relationships. When I say modern society is a de-animating culture, it is to this process that I am referring.

What the transcendence of valuation into a universal equivalent highlights for Marx, is the fracturing of commodities into embodiments of use-value and exchange-value, and the dual nature of labour in its concrete and abstract forms. This idea of fracture or dislocation is central to the subversion of human relations by commodity relations. The evolving social reality of the world of commodities marks the transformation of the "intricate mutual dependencies" of use-values produced by concrete labour through individual effort or reciprocal arrangement, to the universalized measure of exchange-value based on socially uniform conceptions of abstract labour. This two-fold aspect of commodities and labour is reflected in the evolution of social arrangements:

> A thing can be a use-value, without having value. This is the case whenever its utility to man is not due to labour. Such are air, virgin soil, natural meadows, etc. . . . Whoever directly satisfies his wants with the produce of his own labour, creates, indeed, use-values, but not commodities. In order to produce the latter, he must not only produce use-values, but use-values for others, social use-values.[17]

Social use-values communicate through commodity relations which are in turn regulated in terms of equivalency.

In contrast to the use-values of a commodity which "serve the conveniences of human life" — as in the creation of local crafts — exchange-value is "the proportion in which values-in-use of one sort are exchanged for those of another sort, a relation in a larger market that is constantly changing with time and place." With regards to exchange-value, there appears on the epistemological landscape a value form that is at the same time "contained in" the commodity, "yet distinguishable from it." This exchange-value form, which is distinguishable from the inherent qualities of the object, must be "expressed in terms of something common to them all, of which thing they represent a greater or lesser quantity." It is this "something common to them all" which is "distinguishable" from the use-value of the commodity. Exchange-value is therefore "characterized by the total abstraction from use-value." Marx then goes on to state that "there is no difference or distinction in things of equal value." Marx concludes that within this form of abstraction, the commodity's "existence as a material object is put out of sight." I would argue that these same processes of invisibility are found in the natural world, where the mutual dependencies of natural communities are replaced by the standardized measurement of specific resources. This is how the modern world understands nature.

A similar process affects labour in that its two-fold aspect is defined by the exchange of commodities. Labour moves from being particularized and distinctive to being part of a unified process across society. In the same way that nature is universalized as a social use-value, human activity too becomes universalized in the context of commodity relations:

> At first sight a commodity presented itself to us as a complex of two things — use-value and exchange-value. Later on, we saw also that labour, too, possesses the same two-fold nature. . . . [T]his point is the pivot on which a clear comprehension of political economy turns. . . .[18]

The properties of the commodity made invisible by exchange-value, are its specific use-value and the human labour that went into creating it. This labour is inherent in all commodities produced for a society which operates in terms of commodity exchange:

> The labour that forms the substance of value, is homogeneous human labour, expenditure of one uniform labour-power. The total labour power of society, which is embodied in the sum total of the values of all commodities produced by that society, counts here as one homogeneous mass of human labour-power, composed though it be of innumerable indivisible units. Each of these units is the same as any other, so far as it has the character of the average labour-power of society. . . .[19]

Labour then, is the universal measure of the productive capacity of a society which resides in its commodities. Labour is also the basis for the comparative value of particular commodities: "Commodities, therefore, in which equal quantities of labour are embodied, or which can be produced in the same time, have the same value." In the same way that the materiality of the commodity was "put out of sight" by its universal equivalent, so human activity is invisible except as a universalized component of commodities, the value of which is not determined by the context of its creation, but by the market relationship of commodities.

The granting of equal rights to human workers engaged in abstract labour can be seen in this context, as the extension to humans of what was already present in the sociality of commodities. In fact, what has passed for changes in human social conditions in the modern period, has been this filtering of the social realities of commodities to suit those who serve the processes of commodification and production, hence the denial of authorial intentionality in literary theory is linked to the more pervasive aspects of alienated labour.

In the Preface to the 1867 English edition of *Capital*, Marx states that his work views "the evolution of the economic formation of society" as a "process of natural history." Although Marx's analysis highlights the alienation of human society resulting from the forces of production, he nonetheless sees this process as operating within "nature." Marx also makes clear that nature is a resource serving this "natural" process:

> Every useful thing, as iron, paper, etc., may be looked at from the two points of view of quality and quantity. It is an assemblage of many properties, and may therefore be of use in various ways. To discover the various uses of things is the work of history.[20]

This use of nature by humans as part of a "natural" process becomes increasingly problematic within contemporary patterns of resource exploitation. Here, "the work of history" seems to be the destruction of natural communities.

Critical to the inclusion of human history into a universal natural history is the role played by human labour:

> [L]abour . . . is a necessary condition, independent of all forms of society, for the existence of the human race; it is an eternal nature-imposed necessity, without which there can be no material exchanges between man and Nature, and therefore no life.[21]

Once again, in the same way that biological production models (such as the stock assessment models in the fishery that are concerned with the "surplus value" of nature that can be exploited annually) are applied to natural communities, human society is presented by Marx as a production model based on labour.

This production model of society allows Marx to merge "life-process" and "material production" in such a way that the "painful" destruction of natural communities becomes part of "natural history":

> The life-process of society, which is based on the process of material production, does not strip off its mystical veil until it is treated as production by freely associated men, and is consciously regulated by them in accordance with a settled plan. This, however, demands for society a certain material ground-work or set of conditions of existence which in their turn are the spontaneous product of a long and painful process of development.[22]

Although Marx can be criticized for not being sensitive to environmental issues, he nonetheless provides profound insights into the way capital has shaped modern perceptions. For example, Marx presents the enigmatic and mysterious character of money and commodities in this manner:

> A commodity is therefore a mysterious thing, simply because in it the social character of men's labour appears to them as an objective character stamped upon the product of that labour; because the relation of the producers to the sum total of their own labour is presented to them as a social relation, existing not between themselves, but between the products of their labour. . . . The Fetishism of commodities has its origin, as the foregoing analysis has already shown, in the peculiar social character of the labour that produces them. . . . As a general rule, articles of utility become commodities, only because they are products of the labour of private individuals or groups of individuals who carry on their work independently of each other. . . . Since the producers do not come into contact with each other until they exchange their products, the specific social character of each producer's labour does not show itself except in the act of exchange. . . the relations connecting the labour of one individual with that of the rest appear, not as direct social relations between individuals at work, but as what they really are, material relations between persons and social relations between things.[23]

This is exactly the same process that Adam Smith outlined in the above quotation with regard to the origin and use of money. But what was for Smith a natural kind of commercialization of human relations is, for Marx, a process cloaked in mystery and alienation.

This inversion of "material relations between humans" and "social relations between things" provides an insight into the typical mind/matter, spirit/body dichotomies which pervade modern thought. Humans and nature are the matter/body which serve the mind/spirit of commodities. It is Marx's presentation of these kinds of inverted mysteries (with regard to the origin and use of money and commodities) which can offer insights into the difficulties of extricating nature from capitalist categories of exploitation. Marx also elucidates the way these "mysteries" generate modern understanding about the relationship between human and natural communities. Central to this understanding is the uninominal mirror through which the equivalent puts the social relations between humans and nature "out of sight."

Simmel and the Objectification of the Money Form

Georg Simmel's work *The Philosophy of Money*, originally published in 1907,[24] discusses the human experience of the metropolis in the context of commodification rather than in terms of the evolution of the means of production. He is therefore more concerned with the demand side of the commodity equation than with the realities of production, which was Marx's main focus. This explains why Simmel does not concern himself either with the evolution of money into capital, nor the effect of this transformation on social relations. Even though Simmel has been criticized for this shortcoming, a case can be made that an analysis of capital would only have strengthened many of his conclusions.

Simmel focuses on the more individualized and experiential aspects of human reality in the context of commodities and production. Within this context, there is an objectification of the subjective, the quantification of the qualitative, and the equalization of what is not equal. For Simmel, the appearance of the money form affected human experience because it created objects of desire in such a way that difficulty and sacrifice were required in order to obtain them. This, in turn, caused both an internal subjective retreat from, and an abstraction of, exterior reality. Even though his analysis of the money form differs from that of Marx, Simmel — like Marx — connects the transformation of nature and human identity by the abstraction of the value form.

If it is accepted that the utilitarian egoism of Neoclassical economics is not the basis for the universal understanding of social exchange and that market economy is not the natural outcome of the order of things, then it is necessary to consider the constellation of contemporaneous beliefs which give meaning to modern market practices. Simmel presents an analysis that attempts to move beyond the idea of money as an economic indicator, to consider it as part of a more fundamental process related to the universalization of particularity. For Simmel, the social forms created by money place humans at a distance from the substance of things so that they speak "as from afar."

Simmel emphasizes the way money transforms the undifferentiated world into an increasingly rationalized and objectified relation between subject and object:

> . . . mental life begins with an undifferentiated state in which the Ego and its objects are not yet distinguished; consciousness is filled with impressions and perceptions while the bearer of these contents has still not detached himself from them. It is as a result of a second stage awareness, a later analysis, that a subject in particular real conditions comes to be distinguished from the content of his consciousness in those conditions.[25]

These changes are initiated by a process in which "subject and object are born in the same act." Simmel writes:

> In desiring what we do not yet own or enjoy, we place the content of our desire outside ourselves. . . value does not originate from the unbroken unity of the moment of enjoyment, but from the separation between the subject and the content of enjoyment as an object that stands opposed to the subject as something desired and only to be obtained by the conquest of distance, obstacles, and difficulties.[26]

Simmel sees the genesis of value in terms of separation and distance where the formation of value develops with the increase in distance between the consumer and the cause of enjoyment. In this individualistic context, it is possible to say that value is the objectification of desire. In a larger sense then, the cultural process transposes the subjective condition of impulse and enjoyment into a valuation of objects at a distance. It follows for Simmel that the culmination of this process of objectification is the creation of a set of relations which produces goods specifically to be quantified in terms of relative value and presented for the consumption of experience:

> The extent to which money becomes absolute for the consciousness of value depends on the major transformation of economic interest from primitive production to industrial enterprise. Modern man and the ancient Greek have such different attitudes toward money largely because formerly it served only consumption whereas now it essentially serves production. This difference is of extreme importance for the teleological role of money which is the true index of the economy as a whole. Formerly, general economic interest was directed much more towards consumption than to production; agricultural production predominated and its simple and traditional stationary technology did not require as much expenditure of economic consciousness as it did constantly changing industry. . . . [T]his led their [the Greeks'] view of life to be one of a limited and finite cosmos and the rejection of infinity — this trait allowed them to recognize the continuity of existence only as something substantial, as resting upon, and realized in, landed property, whereas the modern view of life rests upon money whose nature is fluctuating and which presents the identity of essence in the greatest and most changing variety of equivalents.[27]

This process occurs only in a pervasive way under the conditions of modern economy. Economic value is therefore the culmination of this process of quantification and objectification, in that money is "the value of things without the things themselves." Value separates itself from the object which it represents. This leads to a situation in which

> the absence of any inherent worth in an object cannot be more distinctly expressed than by substituting for it, without any sense of inadequacy, a money equivalent. Money is not only the absolutely interchangeable object, each quantity of which can be replaced without distinction by any other; it is, so to speak,

> interchangeability personified. The two poles between which all values stand are: at one extreme, the absolute individual value . . .; at the other extreme, that which is clearly interchangeable.[28]

The widespread use of money leads to a situation in which objects become nothing more than "embodied relativity." Money is the symbol of distance, a quantification of loss. It is "nothing but the pure form of exchangeability."

Simmel establishes money as the quantification of the relationship between subject and object. This takes on a pervasive significance under the conditions of modern society as money becomes the means for attaining the object, as well as measuring its value. Money is the means of traversing the distance between subject and object which, in the modern context, is a distance which is never really crossed. In fact, one could say that modern industrial society is always traveling this distance (by monetary means) toward the object of desire but never arriving at its destiny. To quote Simmel:

> . . . [T]he demand for luxury goods is unlimited. . . . This is the result of their superfluousness. The closer values stand to our basic life and the more they are conditions of bare survival the stronger, but also the more limited as to quantity, is the direct demand and the more likely it is that the point of satiety is reached at an earlier stage. On the other hand, the more removed values are from primary needs, the less is their demand measured in terms of a natural need and the more they continue to exist relatively unchanged with regard to their available quantity.[29]

There are similarities between the above quotation and Marx's presentation of the two-fold aspects of the commodity: use-value and exchange-value. But rather than relating this to production as Marx does, Simmel focuses on the subjective experience of consumption. In this sense, the whole development of objects from utility value to aesthetic value becomes a process of objectification. Central to this process is the evolution of money not only as "the purest form of the tool," but also, as the most powerful technology of all. To quote Simmel, "Money is the purest reification of means, a concrete instrument which is absolutely identical with its abstract concept; it is a pure instrument."[30] This pure instrumentality of money is part of a wider social process that infuses an entire society with instrumentality:

> Money is nothing but the vehicle for a movement in which everything else that is not in motion is completely extinguished. . . it lives in continuous self-alienation from any given point and thus forms the counterpoint and direct negation of all being in itself.[31]

For Simmel as for Marx, the self-alienation and negation of being in which meaning is always elsewhere, is demonstrated in the connections between the abstracting aspect of money and the increasingly universal aspect of deities. Both of these processes become increasingly abstracted and universalized in the context of a money economy. Universal money and a universal god promote the same damnation of being. As Simmel states:

> In reality, money in its psychological form, as the absolute means and thus as the unifying point of innumerable sequences of purposes, possesses a significant relationship to the notion of God. . . . In so far as money becomes the absolutely commensurate expression and equivalent of all values, it rises to abstract heights way above the whole broad diversity of objects; it becomes the centre in which the most opposed, the most estranged and the most distant things find their common denominator and come into contact with one another.[32]

The fact that money gives "commensurate expression" to all value, creates a language through which all the world can be translated and converted into economic value. This is what gives money access to so many different communities — and like a disease that is beyond any immune system — is effortlessly predatory.

The monotheistic and abstracted god of Protestantism — as an expression of a particular social form — is, for Marx and Simmel, an expression of the sociality of commodities. Marx states that

> The religious world is but the reflex of the real world . . . And for a society based upon the production of commodities, in which the producers in general enter into social relations with one another by treating their products as commodities and values, whereby they reduce their individual private labour to a standard of homogeneous human labour — for such a society, Christianity with its cultus (sic) of abstract man, more specifically in its bourgeois developments, Protestantism, Deism, etc., is the most fitting form of religion.[33]

For both Marx and Simmel, the particularity of "mutual dependencies" and "distinctiveness" has been absorbed by an extended and homogeneous market of souls.

In terms of abstraction, the money economy created the individual through the intensification of the division of labour and the creation of subjectivity, while at the same time expanding money's capacity to create its own sociality and community. For Simmel, money completes in the economic sphere the general sociological process related to the expansion of the group and the development of the individual. This fracturing continues until the most fully developed money economy (namely, one in which money "works" as with capital) operates according to forces and norms that are relatively independent of those of the owner. Once again, there is an implication that modern sociality exists between commodities which the human owner serves.

For Simmel, when money becomes stored-up means capable of stemming future uncertainty, the fruitful moment has come to a halt in psychological terms, and "whoever serves money is his slave's slave." As the slave of money, Simmel explains the Tragic distance which separates humans from "aliveness," a distance which is measured in terms of money:

> Money is most important in illustrating the senselessness and the consequences of the teleological dislocation, partly because of the passion with which it is craved for, and partly because of its own emptiness and merely transitional character. However, in this sense, money is only the highest point on a scale of all these [modern urban] phenomena. It carries out the function of imposing a distance between ourselves

> and our purposes in the same manner as other technical mediating elements, but does it more purely and completely.[34]

Although Simmel never discusses the relationship between money and nature directly, his use of such terms as "the fruitful moment" and "teleological dislocation," infer that this animated sense of fruitfulness, which he describes as preceding the fracture of subject and object, is related to a sociality of natural communities. Modern uses of money are the most significant causes of this fracturing. This has led to the objectification of nature where money "as the value of things without the things themselves," is a "vehicle for the movement in which everything else that is not in motion is completely extinguished."

Karl Polanyi and General Purpose Money

The "irresistible" forces which the Classical school of economics saw at work in the development of the modern money form, have been presented by Karl Polanyi (as well as others) in terms of fracture and dislocation. By examining non-capitalist sociality and forms of exchange, Polanyi develops a critique which counters theories and practices that conceive of human relations within an industrial economy as the sole basis for understanding all cultures. Polanyi argues that humans do "not act to safeguard [their] individual interest in the possession of material goods; [they] act to safeguard social standing, social claims, social assets. [They] value material goods only in so far as they serve this end."[35]

Polanyi examines this transference of market realities into other cultures, by differentiating between a formal definition of economics and a substantive definition of economics. For Polanyi:

> The formal definition of economics derives from the logical character of the means-ends relationship, as apparent in such words as "economical" and "economizing." It refers to a definite situation of choice, namely, that between the different uses of means induced by an insufficiency of those means. If we call the rules governing choice of means the logic of rational action, then we may denote this variant of logic, with an improvised term, as formal economics.[36]

In contrast, Polanyi defines the substantive meaning of economics as deriving from ". . . man's dependence for his living upon nature and his fellows."[37] This substantive definition refers to the interchange which happens between the natural and social environment. Polanyi outlines the difference between these two definitions in this way:

> The two meanings of economic, the substantive and the formal, have nothing in common. The latter derives from logic, the former from fact. The formal meaning implies a set of rules referring to choice between the alternative uses of insufficient means. The substantive meaning implies neither choice nor insufficiency of means; man's livelihood may or may not involve the necessity of choice and, if choice there be, it need not be induced by the limiting effect of a "scarcity" of the means; indeed, some of the most important physical and social conditions of livelihood such as the availability of air and water or a loving mother's devotion to her infant are not, as a rule, so limiting. . . . The laws of the one [formal] are those of the mind; the laws of the other [substantive] are those of nature.[38]

Polanyi maintains that almost all economic analysis merges these two meanings in a universalizing fashion when in fact the formal definition only applies to sociality under market conditions. As he states, "as long as the economy was controlled by such a [market] system, the formal and the substantive meanings would in practice coincide." Problems would arise, however, when the merged definition was applied to an analysis of cultures that did not operate in the context of market exchange. This transplanting of market assumptions into non-market cultures universalizes market realities in a way that sees all societies behaving in an "economizing" fashion.

Polanyi's contribution to the social form of exchange is not to see it in developmental terms as Neoclassical economics does, but in terms of — to use Ong's[39] words — disruption and contradiction. As Polanyi points out, the developmental perspective of history arises by applying present conditions backwards onto history. This grants history a kind of inevitability and the present a certain objectivity. Or, to reiterate, this kind of project assumes the very processes that need to be explained with regard to the modern destruction of the natural world.

When the substantive definition of economics is extricated from the formal one, the universalizing tendency of modern perspectives is called into question and other forms of exchange besides the modern form of market exchange can appear. Polanyi's extensive analysis of a range of cultures in different historical periods leads him to formulate four forms of integration upon which social exchange is based:

> * Reciprocity: This form operates mainly within the context of the sexual organization of society, "that is family and kinship." Or as George Dalton states ". . . material gift and counter gift-giving induced by social obligation. . . ."[40] These are symmetrical relations existing in an non-hierarchical context — *egalitarian sociality.*
>
> * Redistribution: This pattern of integration is "mainly effective in respect to all those who are under a common chief" and involves ". . . the channeling upward of goods or services to socially determined allocative centers. . . "[41] — *stratified sociality.*
>
> * Household: In contrast to the increasing presence of markets, the household economy is subsistence-based and self-sufficient. As Humphreys states in "History, Economics, and Anthropology: The Work of Karl Polanyi," this category was problematic because "it always applies to a group smaller than society" and was "in any case a vague term defined mainly by an absence of inter-group relations"[42] — *fragmented sociality.*
>
> * Markets: industrial production and market organization create a set of social relations in which, to quote Polanyi, "instead of the economic system being embedded in social relationships, these relationships were now embedded in the economic system"[43] — *dislocated sociality.*

These various forms of exchange can exist side by side at any given time in a society. For example, some of the oldest human cultures had markets just as reciprocity survives in various forms in modern market societies. The pervasiveness of markets in modern society has tended to universalize formal economic theories based purely on markets.

What tends to grant a kind of inevitability to markets is the predatory way in which market realities undermine other patterns of integration. In a discussion about the relations between the traditional society of the Kanak — in what is now New Caledonia — and the French Colonials, Dominique Temple states that market exchange undermines reciprocity in this manner:

> If one puts together a system of reciprocity and an exchange system where both know no other law but their own, so that the exchangers think they are dealing with other exchangers, and donors dealing with other donors, an accretion takes place between the two systems, but they accrete in favour of the exchange system and its triumph. . . . In that sense, the triumph of free exchange is due to circumstance: the defeat of Kanak reciprocity, like that of the Mexicans, the Peruvians and all societies based on reciprocity — India or China — is of a logical order, but this logical relationship is itself possible only under certain precise historical conditions. . . .[44]

Central to Polanyi's discussion about the difference between the various patterns of integration are the terms "embedded" and "disembedded." It is only under the pervasiveness of the market system, as opposed to systems based on reciprocity, redistribution and the household, that social relations become disembedded by relations of commodities and large markets. Instead of economy being embedded in social relations — as they are with reciprocity, redistribution, and households — social relations under markets are transformed into commodity relations, or are disembedded. Within the disembedded framework of a market economy, ". . . man and nature. . . must be subject to supply and demand, that is, to be dealt with as commodities, as goods for sale."[45] Polanyi refers to this process as the fictitious commodification of land and labour. Polanyi describes the ramifications of this transformation on human activity:

> To separate labor from other activities of life and to subject it to the laws of the market was to annihilate

> all organic forms of existence and to replace them by a different type of organization, an atomistic and individualistic one.[46]

This process had a similar annihilating effect on the natural world, both in conceptual and in real terms.

Moving from a formal definition to a substantive definition of economics allows Polanyi to present various patterns of integration based on differing exchange relations. Further extricating the understanding of exchange from market biases, Polanyi defines exchange, not in terms of prices, but as an instituted process which organizes forms of social exchange without any consideration for quantified value. The economy as instituted process refers to the mechanical, biological, and psychological interaction of elements in an institutional frame of reference which in turn gives that process unity and stability. Seen in this way (as opposed to seeing it in terms of a free market), the contrasts between various forms of human and natural communities become evident.

In the same way that he presents various forms of sociality which contrast to the pervasiveness of modern market exchange, Polanyi also presents other forms of money which, while operating within these different patterns of sociality, contrast with the universalizing quality of modern "general purpose" money. This general purpose money is usually regarded as fulfilling four functions: a means of exchange, a means for making (one-way) payments to discharge formal obligations, a store of wealth, and a standard of value. In other cultures, there are instances of local money being used for a specific purpose or to purchase a specific good. These kinds of associations between other forms of integration and special purpose money disappear if general purpose money and markets are universalized. In their article "For a Second Economy," Rotstein and Duncan are critical of the universalizing of modern economic perspectives:

> Many economists are apt to view our modern type of general purpose money as some final pinnacle of progress. While such an evolutionary perspective is rarely acknowledged, it is an underlying leitmotif in modern economic thinking.[47]

In contrast to the assumption that society inevitably moves toward the efficiency associated with general purpose money, Polanyi states that "the substantive definition of money, like that of trade, is independent of markets. It is derived from definite uses to which quantifiable objects are put. These uses are payment, standard, and exchange."[48]

Exchange in unstratified tribal societies based on a substantive definition of money, appears in the form of payment (as it is related to bride-price, blood-money, and fines). In archaic society, these dischargements of obligations through payment, are expanded to include tribute and customary dues. The standardization of goods so that staples may be exchanged is the second form given by the substantive definition of money. The third form concerns the use of money which arises out of a need to quantify objects for indirect exchange. This enables a person to acquire the desired objects through a further act of exchange. Although it is from the perspective of this exchange form that modern society relates to the operation of money, Polanyi states that "in the absence of markets, the exchange use of money is no more than a subordinate culture trait."[49]

What can be concluded from this concept of money is that early money is special purpose money. Different kinds of objects are employed in the different money uses and these uses are instituted differently from one another. Polanyi concludes that money uses — like trade activities — can reach an almost unlimited level of development, not only outside of market-dominated economies, but also in the very absence of markets. As is evident in the above discussion of the way markets undermine other forms of integration, modern general purpose money also undermines the use of special purpose money, which has led some to conclude that there is a kind of irresistible inevitability to the historical movement toward markets. In an analysis of the various special uses of money in the African Tiv society, Paul Bohannan describes different types of money used for specific purposes and for exchanging specific goods. The introduction of modern general purpose money in the Colonial period in West Africa

> provides a common denominator among all the spheres, thus making the commodities within each expressible in terms of a single standard and hence immediately exchangeable [and]. . . has broken down the major distinctions among the spheres. [General purpose] Money has created in Tivland a unicentric economy.[50]

Like the merging of formal and substantive economics under the influence of market societies, special purpose money that exists outside markets, has been integrated with modern general purpose money and hence, has clouded an analysis of the role of exchange in its different forms.

It is interesting to note that in *Capital*, Marx also identifies these three forms of money: payment, standard, and exchange. But instead of seeing them as aspects of use informed by a particular instituted process, Marx presents them in evolutionary terms, as forms leading to the developmental stages of the general equivalent and the commodification of value. From the Classical perspective, these are stages which represent the irresistible movement toward the increased efficiency of formal economic behaviour. For Polanyi, it is simply a fact, not an inevitability, that as a result of particular historical reasons, one form of integration — namely markets — have overrun other forms of integration in the modern period.

What Polanyi challenges more than anything else is the "production model" view of society that has accompanied market perspectives on human activity. Within the discipline of biology, production models of natural communities which focus on achieving the maximum sustainable yield, have long been criticized as reductionist oversimplifications. But this production model still holds sway whenever the "health" of the modern economy is discussed in terms of growth. Surplus economies are supposed to generate utilitarian behaviour amongst its units as each unit scurries around trying to maximize its own self-interest.

This breaking of the mirror of production and the ensuing reconceptualization of social relations based on various forms of exchange, are emphasized by Harry Pearson in his article "The Economy Has No Surplus: Critique of a Theory of Development." Pearson contends that this production model view of the role of surpluses only "arises out of that ideal and institutional complex which views man as economizing atom with a propensity to truck barter and exchange."[51] A stock market, for example, is mistaken for social relations.

In the same way that nature is seen in production models as a kind of factory which generates an annual surplus available for human exploitation, so human society is geared towards generating an annual surplus of production. In outlining alternate forms of integration, Polanyi does not present evolutionary stages which reentrench present market realities, but instead, extricates modern understanding from markets in a way that opens up alternate understandings of human sociality. Polanyi states that, "the market cannot be superseded as a general frame of reference unless the social sciences succeed in developing a wider frame of reference to which the market itself is referable."[52]

Modern political economy is not only a way of doing business and organizing society, it is also a way of understanding social relations. In the context of the discussions that are now taking place with regard to the global environmental crisis, what is required more than anything else is a perspective that is not captive to development, and does not accept as given "the demands of the market." Polanyi's conclusions point toward an understanding of current issues that does not automatically confirm the quasi-objectivist categories of industrial production.

Any meaningful critique of modern economy that is concerned with natural communities must highlight, as Polanyi does in his analysis of archaic societies, the "fictitiousness" of the idea that human identity is based on being operands of production and consumption, and that nature exists only to provide the resources to serve that mechanism. Much of the contentions between Polanyi and his critics are fought back and forth across interpretations of non-market societies. In the same way that Polanyi looks at archaic societies to better understand modern realities and to identify their problems in the context of a global depression and savage world war, can that inquiry not be extended to look at natural communities in a way that extricates them from the resourcist frame of reference? This process can only begin when it is acknowledged that nature is a social place.

Psychoanalysis and Money

In the preface to *Life Against Death: The Psychoanalytical Meaning of History*, Norman O. Brown gives his reasons for pursuing his study of Freud:

> Those of us who are temperamentally incapable of embracing the politics of sin, cynicism, and despair have been compelled to re-examine the classic assumptions about the nature of politics and the political character of human nature. . . .
>
> But why Freud? It is a shattering experience

> for anyone seriously committed to the Western traditions of morality and rationality to take a steadfast, unflinching look at what Freud had to say. . . .
>
> But to what end?. . . It begins to be apparent that mankind, in all its restless striving and progress, has no idea of what it really wants. . . . It also begins to be apparent that mankind, unconscious of its real desires and therefore unable to obtain satisfaction, is hostile to life and ready to destroy itself.[53]

In response to this predicament, Brown calls for a synthesis of psychoanalysis, anthropology, and history. As well as confronting classic assumptions about human character and society, Brown is also responding to neo-Freudians who connect psychoanalytical categories and sociohistorical categories in developmental terms. As such, psychology became abstracted in its representations of an "autonomous soul" developing through history as from childhood to adulthood. For Brown, this undermined the "biological orientation" of Freud's work and its attempt to examine human nature in such a way that shed light on the universalizing and abstracting tendencies of modernity.

What Brown focuses on is the way in which surplus production is dealt with by the social forms of archaic society versus the way a surplus operates in a modern market economy. In historical terms, he sees a close association between surplus production and the sacred, where the surplus is given as a gift to the gods. It is through this ritual of giving gifts that archaic culture attempted to dissolve the beginnings of a division of labour and the growth of a more complex social organization. To quote Brown:

> In the archaic consciousness the sense of indebtedness exists together with the illusion that the debt is payable; the gods exist to make the debt payable. Hence the archaic economy is embedded in religion, limited by the religious framework, and mitigated by the consolations of religion — above all, the removal of indebtedness and guilt.[54]

Brown describes guilt as "aggression towards someone you simultaneously love." This sense of guilt is especially important in terms of familial dynamics where the omnipotent feelings of the child come up against the recognition of powerlessness, and the fact that humans did not indeed make themselves and are not responsible for their existence. A prolonged childhood, combined with the increased rigor of individual self-control in modern society, tends to increase the conflictive feelings which create guilt.

Gift giving ceremonies like the potlatch of the native people in the North American Pacific Northwest, are central to archaic cultures and are therefore central to the response to guilt. There is a deep sense of expiation involved in the archaic notion of giving. This is also the thread that wove the culture together. Reciprocity in archaic cultures is an attempt to share the burden of human existence through an exchange of gifts which have been generated by a surplus economy. This sharing and giving is a very different kind of response to a surplus than is generally associated with modern economizing. It is also a very different response to guilt as compared to modern competitive individualism.

This close association between gifts and the sacred, between economics and religion, existed in Western culture up until the beginning of the modern or Protestant era. Within an increasingly complex social and economic reality brought on by the culture of improvement and the intensification of the division of labour and production, there is an increased sense of guilt. More specifically, there is a recognition that this debt of guilt is no longer payable to the gods through gifts and sacrifice. This is the point at which the gods retreat from the world into invisibility, no longer present to receive the gifts as penance. Brown describes this retreat in terms of the new and more distant Protestant god who no longer receives penance or indulgences and only grants grace arbitrarily.[55] Economics separates from religion as the sin of usury is replaced by money that breeds.

Thus, the increased sense of guilt brings about the emancipation of the economic process from divine control. The secularization of the economy means the abandonment of the comfortable illusion that work achieves redemption, and that the surplus of one's labour can be given to the gods to reestablish a lost balance. It is no accident that the story of Faust — who entered into a relationship with the devil because of disaffection with the world — appeared at this time. He represents the new character archetype who accepted guilt and damnation in his quest for the riches of the world.

In this context, monetary exchange becomes the form of value in which the legacy of reciprocal and redemptive social forms are rendered invisible. To quote Brown, "money is human guilt with the dross refined away until it is a pure crystal of self-punishment."[56] It is a kind of alchemy in which previous social forms are the impure tailings that are cast aside. Monetary wealth in the modern context becomes gathered in death.

In order to articulate this transition to modernity, Brown focuses on the character of Martin Luther and his role in the Protestant Reformation.[57] Brown portrays Luther's recognition of the unredemptiveness of the modern world in which it is impossible to overcome sin through good works. Luther's crisis in faith results from the recognition of a damned world and the consequent retreat of a god from that world. As Luther states, "the devil is the lord of the world." Luther's notion of capitalism as the work of the devil, still admits the problem of guilt into consciousness. It still recognizes the irrational component of these new economic realities, if only through the distortions of religion. Secular rationalism, which also arose during this period and which Luther saw as the devil's "bride and whore," denies the existence of the devil and the damnation of "self-punishment." This denial makes no difference to the economy, which remains driven by this sense of unredemptive guilt. It does however make this difference: that the economy is more uncontrollably driven by the sense of guilt because its true nature disappears from consciousness.

It is interesting to note that not only is the devil central to the economic transformations of Early Modern Europe as in Luther's analysis, but also in the dislocation of pre-capitalist peasants who enter the commodity market as tin miners, as it is described by Michael Taussig. Caught between two worlds, these peasants create gift exchange rituals with the devil world of capital in an attempt to comprehend their situation, while at the same time "standing between the devil and the state, the miners mediate this transformation. . . . [The ritual enactment] is based on the transformation of reciprocity into commodity exchange."[58]

While the rituals of archaic society are based on undoing and expiation, those of modern society are based on sublimation and the displacement of meaning. The line between necessity and superfluity is blurred. A life of chronic non-enjoyment and well scrubbed unhappiness is justified by the hoarding of objects as a newly constellated response to surplus.

Modern economic theory, which accepts as given the demands that appear on the market, also accepts the irrationality of modern human demands for dead things. What the theories of supply and demand really describe are the antics of a sad mammal foraging along store aisles, all the time converting the worthless into the priceless. What makes these kinds of perceptions appear as "eccentric" in the modern social form "is the close connection between money and rationality":

> The connection between money thinking and rational thinking is so deeply ingrained in our practical lives that it seems impossible to question it; our practical experience is articulated in one whole school of economic theorists who define economics as the "science which studies human behaviour as a relationship between ends and scarce means which have alternative uses." The disposal of scarce means among competing ends — what could be more rational than that?[59]

For Brown, this apparent rationality is reconceptualized in psychoanalytic terms as being based upon

> . . . possessive mastery over nature and rigorously economical thinking [that] are partial impulses in the human being which in modern civilization have become tyrant organizers of the whole of human life.[60]

Within the modern social form then, the desire for money takes the place of all genuinely human needs. The effect is a society which substitutes an abstraction, *Homo economicus*, for the concrete totality of human nature, "[a]nd this dehumanized human nature produces an inhuman consciousness, whose only currency is abstractions divorced from real life."[61] Brown also connects the abstractions of currency with the objectification of life as represented in modern science. He sees this modern complex as heir to, and a substitute for, the religious complex which attempts to find god in the world of things. In psychoanalytic terms, modern money — like archaic exchange — is both irrational and sacred, but substitutive abstraction has made this connection invisible.

In attempting to relate these psychoanalytic insights to transformations in social forms of exchange, it is necessary to see money as an abstract replacement operating under the guise of modern market rationality which in turn makes invisible all other forms of human sociality. By extension, it is also possible to state that modern versions of the money form also render invisible previous social relationships between human society and the natural world. Brown outlines the processes of this transformation:

> Whereas for archaic man the crucial defense mechanism is undoing (expiation), for civilized man the crucial defense mechanism is sublimation. . . . New objects must substitute for the human body, and there is no sublimation without the projection of the human body into things; the dehumanization of man is his alienation of his own body. He thus acquires a soul (the higher spirituality of sublimation), but the soul is located in things. Money is "the world's soul."[62]

The insight that psychoanalysis offers to a human society trapped in this "culture of substitution," is that "life is of the body and only life creates values; all values are bodily values."[63]

It is interesting to view economic's disengagement with religion at the beginning of the modern period, in light of attempts to re-engage economics with the environment at the end of the modern period. Unfortunately, in most discussions about sustainable development, it is only the sublimated and reductive forms of *Homo Economicus* and "nature as resource" that become visible as representations of modern sociality. By contrast, what may be required is a project that recreates the symmetrical relationship of reciprocal sociality between humans and nature as the basis for environmental practices. Here, human society gives back to nature its aliveness in order to absolve the guilt inherent in a culture of substitution, and to repay the debt of non-aliveness it has accumulated.

Conclusion

> Economic exchange creates value. Value is embodied in commodities that are exchanged. Focusing on the things that are exchanged, rather than simply on the forms or functions of exchange, makes it possible to argue that what creates the link between exchange and value is politics, construed broadly. . . . This . . . justifies the conceit that commodities, like persons, have social lives.
>
> *Arjun Appadurai*[64]

What is to be made of these disparate yet connected representations about both the distance between tribal and archaic social forms of exchange, and modern market forms of monetary valuation? How do they contribute to an analysis of the relationship between money in its various social forms, and the ways in which modern forms of human sociality at once lead to the destruction of the natural world, and inhibit attempts to develop initiatives and perspectives which would promote the preservation of the few remaining remnants of natural communities?

Modern transformations in social forms of exchange provide an evocative basis for an elaboration of these questions. By extending Polanyi's definition of "disembeddedness" to include not only the extrication of economic practice from social institutions that occurred with the expansion of markets, but also the abstraction of value from the context of the things themselves into a generalized universal form, we can then begin to shed light on the seeming intractability of current environmental problems, such as the collapse of Canada's East Coast fishery.

This disembeddedness refers then to the dislocation of human sociality by the sociality of commodities (in the context of markets), and to the displacement of meaning from the embedded relationships of the things themselves into an abstract quantitativeness as it relates to the exchangeability of commodities. As Harries-Jones, Rotstein, and Timmerman state with regard to the "universal use of money as a mode of communication":

> It provided the common denominator for everything. Moreover, in principle all commodities were transferable and substitutable, one for the other, based on their relative productivity; they could (in theory) be combined in any proportions to maximize some total function such as productivity or profit. In short, the veneer of money permitted precise quantification, relative valuation and substitution of one thing for another, and easy manipulation towards the all-embracing goal of profitable output. Yet, what seems so natural and indeed compelling in a monetized economy dominated by markets, is, by contrast, highly artificial or unnatural in an ecological system.[65]

What this quotation highlights is both Polanyi and Marx's contention that, in a system dominated by markets, it is the commodities that have social relationships which humans and nature serve. Market sociality is commodity sociality. Humans and nature participate in that system to the extent that they are commodified. As Polanyi contends in his description of the commodity fictitiousness of labour and land, "man and nature. . . must be subject to supply and demand, that is, be dealt with as commodities, as goods produced for sale."[66]

Similarly, Marx describes commodities in the context of a universal equivalent, as "citizens of the world" where humans have material relationships and things have social relationships. Within this market sociality, value manifests itself in the social relation of commodity to commodity and, in doing so, the materiality of the "object is put out of sight." So, when I speak of an equivalent which is divided from the thing itself, what I mean by the thing itself is not an essence, but rather a set of embedded social relations between humans and nature which are put out of sight.

This condition of disembeddedness is also central to Simmel's assertion that "value does not originate from the unbroken moment of enjoyment" but from separation of subject and object into a "second stage awareness" which develops with the increase in distance between the consumer and the cause of his enjoyment. For Simmel, money is the most highly developed force causing this separation, and is "the purest reification of means" which in modern society becomes an aimless end in itself. Simmel, too, refers to this process as leading to the extinguishing of the particular within a universal grid of equivalence.

For Polanyi, this disembeddedness occurs when humans and nature become part of the commodification process and "the dislocation caused by such devices must disjoint man's relationships and threaten his natural habitat with annihilation."[67] Central to Polanyi's articulation of this disembeddedness and the way it affects our understanding of the world, is the association he makes between formal economics, "logic" and "mind," and a definition of substantive economics based on "fact" and "nature." I think it is possible to (a) begin with Polanyi's "logic" and "mind," (b) associate that formal definition with the dislocating of value evidenced in the social life of commodities, (c) relate that transformation to the valuation of those commodities by a universal equivalent such as gold, and (d) state that the abstraction of value mirrors the commodity process within the ascendant period of markets in the modern period.

If my contention that since valuation associated with the money form is a significant factor influencing the way the world is understood in the context of markets, then there must be a consideration of the pervasiveness of this process in modern understandings of the world. From the perspective of the preservation of the natural world, the concept of "value" itself has to be questioned. In the context of the discussion here, one could say that what had occurred in the ascendancy of markets is the intensification of valuation, which in turn can be identified as a central impediment to understanding the state of current environmental problems.

I have referred to the modern period as one that has a "dislocating trajectory" associated with the separation of church from state, economics from religion, the individual from the corporate body, god from the Earth, rationality from superstition, the sign from the signified, and nature from culture. I have characterized this separation as the transformation from embedded to disembedded relations. A great deal of modern thinking, at least until recently, has struggled to traverse this separateness. This is the main project of environmentalism. With the increasing pervasiveness of substitutive forms, it may be one of the last fora that struggles to do so.

NOTES

1. Aihwa Ong. 1987. *Spirits of Resistance and Capitalist Discipline*. Albany: State University of New York Press, p. 3.
2. Jean-Joseph Goux. 1990. *Symbolic Economies*. Ithaca: Cornell University Press, p. 97.
3. Thomas Guggenheim. 1989. *Pre-capitalist Monetary Theory*. New York: Pinter, p. 19.
4. Adam Smith. 1961. *The Wealth of Nations*. Indianapolis: Bobbs-Merrill, p. 17.
5. Smith (1961:23).
6. Jurg Niehans. 1978. *The Theory of Money*. Baltimore: Johns Hopkins University Press, p. 1.
7. T. Crump. 1981. *The Phenomenon of Money*. Boston: Routledge, p. 2.
8. Daniel de Coppet (Quoted in Crump.1981:19).
9. B. Malinoski. 1920. "*Kula*: The Circulating Exchange of Valuables in the Archipelagoes of Eastern New Guinea." in *Man*, No. 51, p. 175.
10. Michael Taussig. 1980. *The Devil and Commodity Fetishism in South America*. Chapel Hill: University of North Carolina Press, p. 37.
11. Karl Polanyi. 1957. *The Great Transformation*. Boston: Beacon Press, p. 262.
12. Marshal Sahlins. 1972. *Stone Age Economics*. Chicago: Aldine, pp. 182-83.
13. Karl Marx. 1959. *Capital*. Moscow: Foreign Language Publishing, p. 67.
14. Marx (1959:48).
15. Goux (1990:16).
16. Marx (1959:69).
17. Marx (1959:40).
18. Marx (1959:41).
19. Marx (1959:39).
20. Marx (1959:35).
21. Marx (1959:42).
22. Marx (1959:80).
23. Marx (1959:72-73).
24. Georg Simmel. 1990. *The Philosophy of Money*. New York: Routledge.
25. Simmel (1990:63).
26. Simmel (1990:66).
27. Simmel (1990:232&234).
28. Simmel (1990:124).
29. Simmel (1990:251).
30. Simmel (1990:211).
31. Simmel (1990:510).
32. Simmel (1990:236).

33. Marx (1959:79).
34. Simmel (1990:484-85).
35. Polanyi (1957:47).
36. Karl Polanyi. 1968."The Economy as Instituted Process," in *Primitive, Archaic, and Modern Economies*, edited by George Dalton. New York: Doubleday Anchor, p. 140.
37. Polanyi (1968:139).
38. Polanyi (1968:140).
39. Ong (1987:3).
40. George Dalton. 1961. "Economic Theory and Primitive Society," in *American Anthropologist*, No. 63, p. 9.
41. Dalton (1961:9).
42. S. C. Humphries. 1969. "History, Economics, and Anthropology: The Work Of Karl Polanyi," in *History and Theory*, Vol. VIII, No. 2, p. 204.
43. Karl Polanyi. 1947. "Our Obsolete Market Mentality," in *Commentary*, Vol. 3, No. 2, p. 114.
44. Dominique Temple. 1988. "Econimicide," in *INTERculture*, Vol. 98, pp. 27-28.
45. Polanyi (1957:130).
46. Polanyi (1957:163).
47. Abraham Rotstein and Colin Duncan. 1991. "For a Second Economy," in *The New Era of Global Competition*, edited by Daniel Drache and Meric S. Gertler. Montreal and Kingston: McGill-Queens University Press, p. 417.
48. Polanyi (1968:166).
49. Polanyi (1968:168).
50. Paul Bohannan. 1959. "The Impact of Money on an African Subsistance Economy," *The Journal of Economic History*, Vol. XIX, No. 4, p.132.
51. Harry W. Pearson. 1965. "The Economy Has No Surplus: Critique of a Theory of Development," in *Trade and Market in the Early Empires*, edited by Karl Polanyi, Conrad M.Arensberg, and Harry W. Pearson. New York: The Free Press, p. 321.
52. Polanyi (1968:174).
53. Norman O. Brown. 1985. *Life Against Death: The Psychoanalytical Meaning of History*. Middletown: Wesleyan University Press, pp. xvii-ix.
54. Brown (1985:271).
55. Brown (1985:214).
56. Brown (1985:266).
57. Brown (1985:202-234).
58. Taussig (1980:224).
59. Brown (1985:235).
60. Brown (1985:236).
61. Brown (1985:238).

62. Brown (1985:281).
63. Brown (1985:393).
64. Arjun Appadurai. 1986. *The Social Life of Things*. Cambridge: Cambridge University Press, p. 3.
65. Peter Harries-Jones, Abraham Rotstein, and Peter Timmerman. 1992. "Nature's Veto: UNCED and the Debate Over the Earth," University of Toronto: Science for Peace, p. 5.
66. Polanyi (1957:130).
67. Polanyi (1957:42).

4
Cultures of Substitution

> . . . [A]cademic study of the specific nature of the material artifact produced in society has been remarkably neglected, and that compared, for example, to the discipline of linguistics, our understanding of material culture is rudimentary in the extreme. This lack of concern with the nature of the artifact appears to have emerged simultaneously with the quantitative rise in the production and mass distribution of material goods.
>
> *Daniel Miller*[1]

> With this passage to the political economy of the sign, it is not a matter of a simple "commercial prostitution" of all values . . . . It is a matter of the passage of all values to exchange-sign value, under the hegemony of the code. That is, of a structure of control and of power much more subtle and more totalitarian than that of exploitation.
>
> *Jean Baudrillard*[2]

The way we understand modern society has enormous ramifications for a discussion about human relationships with the natural world. I have described the transformation of modern society as it is absorbed by the imperatives of capital, in terms of the disembedding of both economic forces from social institutions and the concept of value from the embedded relations of the things themselves. I am concerned now with the more pervasive expression of this process in terms of a culture of substitution or a political economy of the sign, in which the substitutive process can no longer be described as being specifically economic, but instead, as constituting a generalized disembeddedness relating to a wide range of activity in modern society. I will discuss conceptions of generalized disembeddedness as represented in Jean-Joseph Goux's *Symbolic Economies* and Jean Baudrillard's *The Mirror of Production*.

When I say these processes are no longer specifically economic, I am not saying that there is not a specifically economic set of processes which continue to drive the expansion of capital. There most definitely is. What has begun to change however, is that while the earlier phases of this process were defined by the separation of specific forms of capital relations from previous forms of social relations, a period of generalized disembeddedness is meant to indicate a situation in which previous forms of relations have been all but absorbed by capital relations in modern Western society. For this reason, a large part of the analysis of the "intellectual ruin" of Western approaches to human-nature relations comes from countries of the South, since many of them have not entirely lost forms of relations which can provide a significant basis for analysing environmental problems.

The constellation of issues that relate to the way society is understood and the effect this has on discussions related to the conservation of the natural world, is evoked in the quotation from Daniel Miller at the beginning of this chapter when he states that "our understanding of material culture is rudimentary in the extreme."[3] Although not concerned directly with the conservation of nature, Miller refers to a transformation in understanding that has occurred in the modern period and which I have discussed in terms of the transformation of the money form into a generalized condition in the political economy of the sign.

Miller's statement highlights the apparent contradiction between the simultaneous "rise in the production and mass distribution of material goods" and the concurrent "lack of concern" with material culture. Instead, modern culture has generated codified disciplines concerned with signs, such as modern linguistics. Miller goes on to state that "material culture is one of the most resistant forms of cultural expression in terms of our attempts to comprehend it."[4] Marx referred to this as the commodity's presence as a material object that is "put out of sight" as it enters the grid of general equivalents.

I find this presentation of current invisibilities in the analyses of material culture to be especially applicable to the representation of the fictitious commodification of labour and land (in the Polanyian sense). What I intend to argue in this chapter is that not only is the modern understanding of material culture "rudimentary in the extreme," but that this is also true in terms of industrial society's inability to recognize the presence of social relationships between human communities and natural communities outside modern categories of economics and abstract-value.

Miller adopts a method which attempts to counteract "highly specialized" and "abstract" philosophical discussions related to an "inauthentic" culture, with "micro-ethnographic" studies of the "particular."[5] This approach is important for demystifying the relationship between society and nature as well. In *The Social Life of Things*, Arjun Appadurai confronts methodological issues in a similar way:

> Even if our own approach to things is conditioned necessarily by the view that things have no meanings apart from those that human transactions, attributions, and motivations endow them with, the anthropological problem is that this formal truth does not illuminate the concrete, historical circulation of things. For that we have to follow the things themselves, for their meanings are inscribed in their forms, their uses, their trajectories. It is only through the analysis of these trajectories that we can interpret the human transactions and calculations that enliven things. Thus, even though from the theoretical point of view human actors encode things with significance, from a methodological point of view it is the things-in-motion that illuminate their human and social context.[6]

Although his representations never move beyond human-centredness, Appadurai's challenge to "the view that things have no meanings apart from human attribution" is endemic to discussions concerning the invisibility of the sociality of nature. In order to illuminate the human and social context in current representations of nature, it is necessary to "enliven" in natural communities the relations of "the things themselves," and use them as a basis for analyzing those contexts. We must also recognize these representations as discursive forms that are normally invisible to human society. This will ultimately cast light on the current "environmental" crisis.

Money That Breeds

> Remember, that money is of the prolific, generating nature. Money can beget money, and its offspring can beget more, and so on.
>
> *Benjamin Franklin*[7]

The history of modern society is Franklin's "and so on." The naturalizing of money as a beneficence — its prolific nature a boon to human society as it creates chapters and verses of begotten off-spring — combined with the blithe acceptance of Franklin's "and so on," embody the main concerns of my inquiry into the destruction of the natural world. Although the ramifications of "and so on" are obviously clearer in the latter twentieth century than they could have been to Franklin in the eighteenth century, many of the root causes of the dark side of "and so on" still remain hidden in assumptions that "naturalize" money and support its begetting.

I will explore the way that new forms of exchange based on capital have transformed general conceptions of valuation from what Benjamin Nelson in *The Idea of Usury,*[8] describes as the particularity of tribal brotherhood, to the "transvaluation of values" into universal otherhood where everyone is a stranger as opposed to the usual tribal exemption. As he states, "in modern capitalism, all are 'brothers' in being equally 'others.'"[9]

Unlike commodities, which find their origin in the materiality of the natural world, capital originates in money. The difference between money and capital "is nothing more than a difference in their form of circulation."[10] The form of circulation which facilitates the exchange of money is expressed by the transformation of commodities into money, which is then changed back again into commodities (C—M—C) — "or selling in order to buy." By contrast, capital circulates by transforming money into commodities, which are changed back into money (M—C—M) — "or buying in order to sell." Marx contrasts these processes:

> In the one case both the starting point and the goal are commodities, in the other they are money. In the first form the movement is brought about by the intervention of money, in the second by that of a commodity.[11]

In other words, money is spent "for the satisfaction of definite wants," while capital is advanced so that the process "begins and ends with the same thing, money, exchange-value; and thereby the movement becomes interminable."[12]

Within the dynamic of circulation outlined by Marx, it is possible to see the way in which capital further encloses human society within the logic of general equivalents. Instead of facilitating the circulation of the money sign which continually refers back to the "satisfaction of definite wants," capital creates a form of exchange in which the profitability of the sign is the sole rationale. The materiality of the world is only seen in terms of the expansion of quantification within the grid of general equivalents. As such, the archaeology of the economic value form is a movement from "the things themselves" in barter, to "an end in itself" in capital.

Capital can be said to "enter into private relations with itself" whereby "it has acquired the occult quality of being able to add value to itself. It brings forth living offspring, or, at the least, lays golden eggs."[13] The expansion of value is the active factor in a process which "presents itself as an independent substance, endowed with a motion of its own, passing through a life process of its own, in which money and commodities are mere forms which it assumes and casts off in turn."[14] Within the context of modern society, there also appears the new form of interest-bearing capital which becomes "value that is greater than itself" without ever taking on the temporary identity of a commodity. This is the naturalization of a dead thing that grows. This is the eugenics project of modernity.

In the last two hundred years of Western history, money's incarnation as capital is probably the most definitive process which has shaped society. To quote Taussig:

> In the realization of this aim [capital accumulation], capitalism stamps its products and its means of production with the seal of market approval — price. Only by "translating" all the varied qualities that constitute its products and means for creating them into one common "language," that of currency, can the generator of capitalist's mentality, the market, operate.[15]

This is the transvaluation which occurs as a result of the relations within capitalism. Robert Heilbroner outlines a process similar to that of Marx:

> This repetitive, expansive process is, to be sure, directed at bringing goods and services into being through the organization of trade and production. But the physical attributes of these commodities, even when they take the form of luxurious objects, are not prized as evidences of a successful completion of the search for wealth, as long as they are in the capital-

> ist's possession. On the contrary, their physical existence is an obstacle that must be overcome by converting the commodities back into money. Even then, when they are sold, the cash in turn is not regarded as the end product of the search but only as a stage in its never-ending cycle [of expansion].[16]

Capital is not a material object, but a process which temporarily resides in material goods, before reconstituting itself as more capital. It is a deadening process which has entered into private relations with itself. "And so on."

Human identification with these processes is what perpetuates them. As Marx states:

> . . . [I]t is only in so far as the appropriation of ever more and more wealth in the abstract becomes the sole motive of his operations, that he functions as a capitalist, that is, as capital personified and endowed with consciousness and a will.[17]

It is precisely this expansive process which regulatory agencies and those who wish to manage development can do nothing about. The essence of capital is expansion: "Without the organizing purpose of expansion, capital dissolves into material building blocks that are necessary but not sufficient to define its life purpose."[18]

The "life purpose" of infinite value leads to the undermining of human communities and natural communities. Just as every soul desires to return to god, it is money's goal to return as more money. Capital's worst nightmare is to be trapped in the particularity of something that cannot be sold. Dead stock. Hell on Earth.

The World Eaters

> Above all it is necessary to understand the way in which the market system of modern capitalism engenders a marketing mentality in which people tend to be seen as commodities and commodities tend to be seen as animated entities that can dominate persons.
>
> *Michael Taussig*[19]

The social ramifications of the expansion of capital can be expressed in the transformative concepts of reification and commodity fetishism. An understanding of these concepts is central to the processes of demystifying capital. It is these transformative concepts which are originally associated with economics specifically, but become a generalized process in cultures of substitution. To quote Taussig again:

> . . . [T]he social process of capital reproduction and expansion may easily appear as a quality inherent in the commodity itself, rather than of the process of which it is a part. This socially conditioned appearance is a mystification in which the entire social context conspires, so to speak, to mask itself. In this process of decontextualization, profit no longer appears to be the result of social relation, but of a *thing*: this is what is meant by *reification*.[20]

Whereas reification operates predominantly within the realm of production, commodity fetishism describes the context of consumption within modern market economy,

> . . . wherein capital and workers' products are spoken of in terms that are used for people and animate beings. It is money as interest-bearing capital that lends itself most readily to this type of fetishism. Capital appears to have the innate property of self-expansion, and this property diffuses into all economic life since in capitalism money is the universal equivalent and mediator between persons and objects. . . . Fetishism denotes the attribution of life, autonomy, power, and even dominance to otherwise inanimate objects and presupposes the draining of these qualities from the human actors who bestow the attribution.[21]

Fredric Jameson outlines a similar transformation of human identity that occurs within modern capitalist society:

> The theory of reification . . . describes the way in which, under capitalism, the older traditional forms of human activity are instrumentally reorganized . . .

> and essentially restructured along the lines of a differentiation between means and ends. . . hence the strategic value of the Frankfort School term "instrumentalization" which usefully foregrounds the organization of the means themselves over against any particular end or value which is assigned to the practice. In traditional activity, in other words, the value of the activity is immanent to it, and qualitatively distinct from other ends or values articulated in other forms of human work or play. . . . It is only with the universal commodification of labor power. . . that all forms of human labor can be separated from their unique qualitative differentiation as distinct types of human activity. . . and all universally ranged under the common denominator of the qualitative, that is, under the universal exchange-value of money.[22]

Jameson also sees the complementary process of commodification as a significant contributor to the transformation of human identity within capitalism:

> The concept of the commodity cuts across the phenomenon of reification — described above in terms of activity and production — from a different angle, that of consumption. In a world in which everything, including labor power, has become a commodity, ends remain no less undifferentiated than in the production schema — they are all rigorously quantified, and have become abstractly comparable through the medium of money, their respective price or wage — yet we can now formulate their instrumentalization, their reorganization along the means/ends split in a new way by saying that, by its transformation into a commodity, a thing of whatever type has been reduced to a means for its own consumption. It no longer has any qualitative value in itself, but only insofar as it can be "used": the various forms of activity lose their immanent intrinsic satisfactions as activities and become means to an end.[23]

The quantitative, universalizing, departicularizing "quality" of expansive capital has a profound effect on social relations, in that the basis for these relationships is defined in terms of a grid of equivalents. This attribution of life and power to otherwise inanimate objects has the effect of denying humans life and power. The transformative concepts of reification and commodity fetishism that are associated with the defining aspects of economic realities (in the context of capital), are central to the disembedding processes that gradually become a generalized condition in cultures of substitution. What these processes represent is a mystification in which the relations of the things themselves disappear, as reality becomes increasingly expressed in terms of categories amenable to capital. In the movement to a culture of substitution, the social relations of commodities becomes the general condition of all activities. What such terms as reification and commodity fetishism convey, is the massive inversion which has occurred in modern society where deadening processes dominate the living.

Cultures of Substitution

> . . . the major scheme of the modern period [is] that it makes human beings increasingly dependent upon totalities and universalities and increasingly independent of particularities.
>
> *Georg Simmel*[24]

Within modern culture, the processes of reification and commodification are indicative of the increasing impoverishment of the social relationship between humans and nature. Ascendant political economy is a record of changing conceptions of a stand-in, given "life-like" qualities which in the modern period gradually severs its involvement with the relations of the things themselves. Goux regards his discussion about cultures of substitution as an extension of this process:

> A fresh reading of the genesis of the money form elaborated by Marx allows us to discern a structural logic of the formation of the general equivalent, a logic that leads to my methodological extension of this notion to other domains, where values are no longer economic, where the play of substitutions defines qualitative values.[25]

Generalizing the idea of exchange in order to understand the symbols through which a society expresses itself, Goux's analysis provides a basis for understanding the frame of reference which defines human monoculture and informs modern discussions about the conservation of the natural world. More than anything else, the logic of unified general equivalents is a rejection of diversity of the Earth. Goux contends:

> Indeed, what the generalized concept of exchange made possible was the definition of major social formations as a *mode of symbolizing* that is both economic and significant. The structural homology among the various registers of exchange could aptly guide an analysis of the historical correlations between particular symbolic institutions.[26]

Through this *mode of symbolizing*, as expressed in the context of exchange, it becomes possible to relate a wide range of societal beliefs and describe them in terms of a culture of substitution. This substitution expresses itself in linguistic, commercial, sexual, and legal terms, as concurrent processes reflecting similar transformations into a grid of general equivalents.

What Goux outlines as a dialectic process that overcomes contradictions in the transition from a natural economy to a capitalist economy, can also be described as the increasingly monolithic nature of substitutive processes. For Goux, this is a logico-historical dialectical transformation which engages all aspects of human relationships and understanding:

> If there is such a relation between logic and history, and if the link between social exchange and the mode of production is indissoluble, the periodization of the emergence of money corresponds to the history of successive principal forms of social exchange, that is, to the sequence of modes of production and exchange in the course of historical development — to their stratified typology. . . . To put it quite baldly, then, I find the following correspondence between the typical historical sequence of modes of production and exchange: *F1, primitive mode of production; F2, Asiatic (or tributary) mode of production; F3, N1, ancient mode of production; N2, feudal mode of production; F4, N3, capi-*

> *talist mode of production*. . . . We can envision not only a logical sequence of exchange forms . . . but also a history of the major social modes of exchange in succession; with these homologies laid bare, this history expands immediately into the history of the *modes of symbolizing*.
>
> I submit that within a given mode of production, the dominant and specific status of exchange, the *form of exchange*, can be observed at numerous levels of the social whole, with the understanding that exchange (substitution, equivalence, metaphor, representation, etc.) occurs at levels as different as kinship, writing, sexuality, religion and other institutions.[27]

As an example of this dialectical process, Goux links the evolution of written language, from pictographic, to ideographic, and to phonographic representation, with the different phases of symbolization whose logical base is furnished by the genesis of the money form (in terms of a transformation from the materiality of barter or pictographs, to the increasing immateriality of signs and sounds).

Central to this transformation of the modes of symbolizing are the changes that occur in human consciousness:

> To show how the very form of social consciousness in a given mode of production and exchange is determined . . . is to become able to consider consciousness (social or individual) no longer as a simple mirror, an unvarying agency of reflection, but rather as *constituted* in its very form, in its *mode of reflection*, by and in the process of social exchange. The "content" is what reflects, to varying degrees, real social existence, whereas the mode of reflection is completely and effectively determined, entirely apart from the subject's consciousness, by his multiple exchange relations and, to be precise, by the stage of the logic of symbolization in which the exchanges are located.[28]

This historically constituted approach to understanding leads to what could be called the archaeology of symbolization:

> Only a theory of social organization which took into account the dialectical process of "symbolization,"

> that is of the *social exchange of vital activities* — going beyond the mechanistic economist dichotomy between material infrastructure and superstructure — could account for this socioeconomic totality or for this stage in the history of *culture*, in the anthropological sense of the word. . . [and]. . . would constitute (as the by-product of a theory of social organization) a *theory of philosophy*, paving the way for another level of gnoseological systematization than what is now still known as "philosophy."[29]

Goux's rendering of the relationship between "vital activities" and the symbolization process is evocative in the context of the relationship between the natural world and human culture. Although Goux's main focus is on the constitution of gender within patriarchy, his presentation of this archaeology of symbolization has significance for discussions about the sociality of nature.

Goux describes idealist philosophy associated with the ascendancy of the modern project as an example of this mode of symbolizing:

> The idealist optical illusion consists in viewing the visible material world as the reflection of general equivalents, whereas general equivalents constitute the focused reflection, the specular image, of the visible world's multiplicity and differentiation.[30]

Of course, multiplicity, differentiation, and particularity are central to arguments for conserving nature. This is also the case Miller makes in his discussions of micro-ethnography, and which are mounted in opposition to the generalizing monocultural mode of the modern value form.

Goux contrasts the symbolization of the modern value form with archaic forms of reciprocal relationships, in terms of the "evocatively charged. . . layers of richly significant symbolism that characterized barter,"[31] or what he refers to as Ferenzi's discussion of "the real symbol" which has an unconscious identification with something else. This is in contrast to modern symbolization in which, "through the withdrawal of primary investments, a conscious, disaffected, depersonalized symbolism is reached."[32] I described this transformation in the previous chapter in terms of Polanyi's disembeddedness as well as Marx's representation of capital, which transforms exchange between the things themselves into exchange as an end in itself. For Goux, this affects the relationship between human society and the natural world:

> What is banished, completely excluded from capitalist sociality is the surplus of meaning arising from an unconscious identification with something else. As a predominant social relation becomes founded on economic surplus value, this relation suffers from a depreciation of meaning. . . [and] the dominant ideology is caught up in an instrumental rationality.[33]

Meaning does not reside in a growing and dying natural community, but in an ever-expanding system of signs. This instrumentality leads to a set of social relations which are a culmination of the process of disembeddedness inside what Goux describes as technological sociality, expressed here in terms of abstract art:

> With abstract painting. . . confidence in poetic generation takes a flying qualitative leap, leading to the denial of the preconstituted soil of existing nature. We see a rejection of any metapictorial guarantee ensuring the meaning and order of the canvas, any centralized reflection of something already there. No longer does a stable, focused aspect of the subject repeat in pictorial space a referential *given* that preordains the painted signifiers; the dislocated subject is invaded, so to speak, by its own operative and generative capacity.[34]

"The denial of the preconstituted soil of existing nature" and the invasion of the "dislocated subject" point to an archaeology of the economic value form which links the environmental crisis with the crisis in human identity. It is clear that this process is no longer just economic and can now be described as a generalized disembeddedness.

Goux articulates the ramification of this change from market sociality to technological sociality as such:

> The market rationality has given way to an epistemo-technological rationality, replacing one of equivalence, value, reflection, reproduction, being, or

> essence with one of operation, which precedes the fixed exchange and the code.[35]

In other words, the symbol does not refer back to anything. At the same time, defining aspects of commercial relations disappear so that "no economic base can any longer be autonomized with respect to symbolic practices." Here then, the defining aspects of modernity based on separation and autonomy which I referred to earlier (church from state, economics from social relations, individual from collective, rationality from belief) come to an end, and in their place appears a system of signs which no longer accommodates sets of relations which allow for the appearance of the relations of the things themselves.

This enclosure of relations within a self-referential operationalism has profound effects on the way technological society understands and relates to nature:

> All the organizational and informational power that is immanent in "nature" — that is, in its generative operation — is denied and negated, only to be chalked up to a reason, to a logos spermaticos, to a separate intelligence with a male charge, while all that remains of this nature is its negativity as passive receptacle, as neutral, plastic substance, reduced to the role of a simple matrix (womb), mater. Thus the idealist notion of matter can be supported only by a conception of matter as reduced, amputated thought. Generative nature is the unthought absolute of an idealism that can conceive only a material nature, for the good reason that thought and consciousness are posited as separate only in place of what they deny nature, its immanent organizational potency.[36]

Goux states that "[this schism] makes genetic continuity between nature's productivity and thought unthinkable. . . and places materiality outside of value and meaning."[37] Goux's representation could be described as the basis upon which humans (assuming nature can't manage itself) presume their qualifications to "manage" nature. It is this archaeology of value which can provide a critique of most of the current discourse on managing the Earth. It is also this archaeology which remains invisible if the mystifications of capital are not questioned.

Despite the fact Goux contends that "no economic base can any longer be autonomized" from other symbolic registers in a technological society, he still recognizes a logico-historical process in which class struggle is a significant exemplar of the symbolization process. Goux also sees this historical process as necessary:

> Patriarchal hegemony is no accident, no mere coincidence, but a necessity of human evolution considered as History, or rather, considered as having traversed a certain mode of *historicity.*"[38]

Because of his interest in the creation of gender, the nature that Goux talks about is feminine nature and not natural community. Also, when he talks about the necessity of history, it is based on the fact that

> The ontogenetic journey from mother to woman requires mediation by a phallic instrument, enabling the male subject to accede to a reality that coincides in principle and position with a reunion with the woman.[39]

This is a resolution of gender in a new dialectical materialism. It is not the "primitive materialism," of prehistory, but "historicized materialism," and thus a historicized nature. Mother Nature must become the bride of modernity.

Goux describes the course of human history in this way:

> Thus the historical process as a whole, in a grandiose and enigmatic loop, would seem to lead from the inclusion of man in nature back to the inclusion of man in nature; but the loop is progressive rather than regressive. For what is rediscovered is an *other* nature, transformed, socialized.[40]

Because Goux reads class struggle and labour into the logico-historical process, the resolution into a new materialism remains within the confines of production and the historicization of nature. In the alchemy of human history becoming natural history, natural communities do not reappear except as part of the historical process. For Goux, the social is enclosed within human culture and, although his argument continually points in its direction, he does not grant nature any social standing.

Despite his acceptance of the ancillary role nature must play in the logico-historical human drama, Goux provides a profound critique of the modes of symbolizing shown by the archaeology of the value form, as it is transformed from a specific association with economics to a generalized condition in technological society. He sees the mirror of production as a historically constituted, but nevertheless a necessary stage in the socialization of humans and nature.

Baudrillard and the Political Economy of the Sign

> The mirror of political economy is . . . the identity that man dons with his own eyes when he can think of himself only as something to produce, to transform, or to bring about as value.
>
> *Jean Baudrillard*[41]

> In order to question the process which submits us to the destiny of political economy and the terrorism of value . . . the concepts of production and labor developed by Marx (not to mention political economy) must be resolved and analysed as ideological concepts interconnected with the general system of value. And in order to find a realm beyond economic value (which is in fact the only revolutionary perspective), then the *mirror of production* in which all Western metaphysics is reflected, must be broken.
>
> *Jean Baudrillard*[42]

Although both Goux and Baudrillard recognize the enclosing qualities of a self-referential code, they differ in terms of their views on whether Marxist analysis is up to the task of being able to describe this self-referentiality. Goux claims that it can, whereas Baudrillard attempts to extricate his analysis from the mirror of production in order to undo any logico-historical misconceptions about the necessary evolution of human society which he sees as being inherent in Marx as well as Neoclassical political economy.

After quoting Marx's statement concerning the significance of utility and production in human history, where "men begin to distinguish themselves from animals as soon as they begin to *produce* their means of subsistence. . . ," Baudrillard asks, "why must man's vocation always be to distinguish himself from animals?"[43] Beginning with this question, Baudrillard sets out to dismantle the mirror of production which has been used to understand modern political economy, while trying to "find a realm beyond economic value" which can provide a basis for analysing the political economy of the sign.

For Baudrillard, both political economy and its critiques reduce human identity and nature to values which require utility and production in order for them to be actualized. This leads to an analysis of human society based on modes of production and relations of production. But Baudrillard argues that there is a second stage in political economy which relates to a process of social abstraction referring not only to the commodity but also to the sign, which Marx did not recognize. To engage in this kind of analysis, Baudrillard employs the concepts of contemporary structural linguistics. To quote Mark Poster:

> . . . Baudrillard argues that the essence of political economy is precisely this separation [signifiers have become abstracted from the subject (the signified) and from the social world of objects (the referent)]; the increasing autonomization of the signifier not simply in the realm of language but in all aspects of social exchange. Marx foresaw that capitalism would corrupt all values, moral, cultural sexual, etc., by the force of the exchange-value of the commodity. Baudrillard asserts that the strategy of the capitalist system is to generate this abstract structure of signification of which the commodity is merely one example.[44]

Similar to Goux's contention that technological society supplants the "real symbol" which is connected to "something else," with a disaffected and depersonalized form of symbolization, Baudrillard argues that the code, in a generalized formulation, no longer refers back to any subjective or objective reality, but is instead trapped within the operation of its own logic. In this context, a generative nature does not appear for Goux because it is the "unthought absolute" while for Baudrillard, nature is the "unmarked term" and "zero point" of the code.

It is where Goux and Baudrillard differ that provides an evocative starting point for a discussion about reconceptualizing forms of exchange that extends analysis to include the sociality of natural communities, and finds "a realm beyond economic value." As stated earlier, Goux begins by applying Marx's concepts of utility and labour as central aspects of his analysis, and ends up with a historicized concept of nature. The importance of this frame of reference for understanding the invisibility of social nature in modern discourse cannot be underestimated. Although Goux makes an immense contribution to discussions about the invisibility of nature, as when he characterizes human history in terms where "it is always through replacement that value is created,"[45] his perspective is limited by his universalization of utility and labour. By universalizing utility and labour as endemic to human identity, he now needs a historicized nature which is part of human history. In Goux's terms, "primitive nature" — and all the buried understandings of nature that are latent in that term — requires human history for it to be transformed into historicized nature. It is not enough to leave nature on its own. This perspective is at the heart of the impoverishment of nature by the terrorism of value.

In contrast to Goux, Baudrillard questions these conceptions of utility and labour, and points to a critique of human history which does not privilege these categories of understanding. By making utility and labour problematic, Baudrillard attempts to extricate the discussion from these categories, as well as assumptions of a *logos* in the development of human culture. This kind of approach would allow for a nature which is outside the resourcist paradigm:

> It is only in the mirror of production and history . . . that our Western culture can reflect itself in the universal as the privileged moment of truth (science) or of revolution (historical materialism). Without this simulation, without this giant reflexivity . . . our society loses all privileges. It would not be any closer to any term of knowledge or any social truth than any other.[46]

In order to make visible the impoverishment involved in a conception of human identity which is linked to utility and labour, and by association, the forces of commodification and production which impoverish nature, Baudrillard presents a critique of the universalization of utility which connects productive relations with logico-historical process.

For Baudrillard then, it is not enough to analyse the operation of the quantitative abstraction of exchange-value starting from use-value. It is also necessary to problematize this stage of use-value which was supposed to precede exchange-value. Instead of contrasting the more authentic and differentiated reality of use-value with the abstracted and alienating abstract-value which forms the basis for Marxist analysis, Baudrillard argues that this binary approach to use-value and abstract-value is precisely what has caused the universalization of utility and efficiency (i.e. value) throughout human and natural history. Any analysis of political economy must then operate within that frame of reference, and must always inevitably lead to a conclusion that nature is necessarily bounded by utility and labour. Baudrillard states:

> What produces the universalization of labor in the eighteenth century and consequently reproduces it is not the reduction of concrete, qualitative labor by abstract, quantitative labor but, from the outset, the structural articulation of the two terms. Work is really universalized at the base of this "fork," not only as market value but as human value. Ideology always thus proceeds by a binary, structural scission, which works here to universalize the dimension of labor. . . quantitative labor spreads throughout the field of possibility. Henceforth there can only be labor — qualitative or quantitative. . . . It signifies the comparability of all human practice in terms of production and labor. . . . In this structuralized play of signifiers, the fetishism of labor and productivity crystallizes.[47]

Echoing Polanyi's analysis in *The Great Transformation*, Baudrillard states that the focus on labour and utility can undermine the recognition of other forms of exchange:

> Assuming the generic schema of production and needs involves an incredible simplification of social exchange by the law of value. . . . The dialectic of production only intensifies the abstractness and separation of political economy.[48]

Science, technique, progress, history — in these ideas an entire civilization based on the "law of value" results in a condition where ". . . the transformation of nature is the occasion of its objectification as a productive force under the sign of utility."[49] Central to Baudrillard's argument is the recognition that the law of value which developed under the sign of utility is a historical contingency rooted in a particular set of social relations. When this contingency is not recognized, there exists a form of analysis which, "as soon as they [value and utility] are constituted as universal they cease to be analytical and the religion of meaning begins."[50]

In order to undermine this "religion of meaning" and break the mirror of production, Baudrillard contends that production and utility cannot provide analytical insight, and that only through a reconceptualization of symbology and forms of exchange can modern political economy be understood. Within the context of generalized disembeddedness, production and utility, like commodification, have an increasingly difficult time providing a form of analysis of this generalized condition.

To provide a frame of reference for analysing this new historical reality, Baudrillard points to a form of symbolic exchange based on a realm of meaning beyond economic value, similar to what Polanyi proposed when he stated that "the market cannot be superseded as a general frame of reference unless the social sciences succeed in developing a wider frame of reference to which the market itself is referable."[51] Like Polanyi, Baudrillard looks to other forms of social exchange which can provide a basis for analysing modern sociality:

> Instead of exporting Marxism and psychoanalysis [in an analysis of primitive society] . . . we bring all the force and questioning of primitive societies to bear on Marxism and psychoanalysis. Perhaps then we will break this fascination, this self-fetishization of Western thought.[52]

I think it is possible to link the arguments that I have been making about the extrication of discussion from the categories of capital, with Baudrillard's analysis which explains the sign's relationship to symbolic exchange. For Baudrillard, as for Polanyi, earlier forms of social exchange — even that of the master-slave — were based on reciprocal social relationships whereby "the collective process remains internal to the group," and where relations are defined "through the reciprocity of the group."

The reciprocity of certain forms of social exchange exists through language and communication, not through utility and production. This concept of exchange through language and communication has a profound effect on the form of analysis, in contrast to political economy analysis which is based on production. As an example of this contrast, Baudrillard states:

> The work of art and to a certain extent the artisanal work bear in them the inscription of the loss of the finality of the subject and the object, the radical compatibility of life and death, the play of an ambivalence that the product of labor as such does not bear since it has inscribed in it only the finality of value.[53]

Within this ambivalent world of contradictions that are expressed in terms of forms of exchange, relations are not synthesized in any dialectic or enclosed by a finality of value. As opposed to the separation and autonomy synonymous with the ascendancy of commodity relations in modernity, and the gradual enclosure endemic to the crisis of modernity, "the symbolic sets up a relation of exchange in which the respective situations cannot be autonomized . . ."[54] In other words, it is capable of defining embedded relations, something market exchange cannot do.

This sociality is a cycle of endless "giving and receiving" based on the "radical compatibility of life and death."[55] It recognizes the multiplicity and interdependency of meaning. It also ensures a "loss of finality" in the individual subject. Here, the mirror of production can be broken open so as to include the social relations of natural communities. It is here that the "unfinalized" identity of nature — to use Thomas Birch's term — can be contrasted to the finality of value. It is through this metaphor of giving and receiving that a more profound analysis of modernity can be made to extend beyond earlier human societies, and acknowledge that nonhuman nature is a profoundly social place.

Within Baudrillard's presentation of the generalized condition of the political economy of the sign, symbolic forms of exchange based on communication rather than production inevitably undergo fracture and dislocation:

> This stage is accompanied by a radical change in the functioning of the sign, in the mode of signification . . . . [T]he signified and the referent are now abol-

> ished to the sole profit of the play of signifiers, of a generalized formalization in which the code no longer refers back to any subjective or objective "reality," but to its own logic. . . . The sign no longer designates anything at all.[56]

Baudrillard refers to this condition as the "domestication of language" in which "the domestication of all social and symbolic relations in the schema of representation, are not only contemporary with political economy, they are its very process . . ."[57] As "the very process" of the political economy of the sign, this condition can also be defined as a culture of substitution. The use of the terms "domestication of language" and "domestication of social relations," are evocative in terms of extending the analytical bases of social forms to include the "giving and receiving" of undomesticated natural communities.

By questioning the logico-historical process operating in terms of utility and production, and the way that this process has become generalized in the political economy of the sign, Baudrillard claims that a new set of relations requiring a different form of critique has emerged:

> The truly capitalist phase of socialization through labour and the intensive mobilization of productive forces has been overturned. . . . Surplus value, profit, exploitation — all these "objective realities" of capital have no doubt worked to mask the immense social domestication, the immense controlled sublimation of the process of production. . . . The system is reproduced today through a reverse tactic: no longer one of general mobilization, but of techno-structural rationalization, that has as its effect the corruption of all categories.[58]

Human history is characterized by Baudrillard as a process of social domestication, and not a logico-historical development. Modern social forms have become increasingly problematic with regard to discussions about creating viable relations *between* humans, as well as *with* the natural world:

> Its [capitalism's] fatal malady is not its incapacity to reproduce itself economically and politically, but its

> incapacity to reproduce itself symbolically. The symbolic social relation is the uninterrupted cycle of giving and receiving . . . . It is this symbolic relation that the political economy model (of capital), whose only process is that of the law of value, hence of appropriation and indefinite accumulation, can no longer produce. It is its radical negation. . . . It is this fatality of symbolic disintegration under the sign of economic rationality that capitalism cannot escape.[59]

This is the engine of generalized disembeddedness which undermines the "uninterrupted cycle of giving and receiving." Brown, for example, discusses the way symbolic formation was transformed in Early Modern Europe, from expiation and giving in order to lose guilt, to formations based on sublimation and denial, with the end result being competitive "rational" individualism.

Capitalism cannot escape from indefinite accumulation while it is enclosed by the code. It is also this process which destroys social relationships:

> It [late capitalism] can no longer achieve any symbolic integration (the reversibility of the process of accumulation in festivals and waste, the reversibility of the process of production in destruction, the reversibility of the process of power in exchange and death.). . . . Everywhere the pressure of the system of political economy is heightened. The final avatars are anti-pollution and job enrichment. Here also the system seems to slacken its limits and restore nature and work to their dignity. . . . But we know very well that a symbolic relation of man to nature or to his work will not reemerge here. There will only be a more flexible and reinforced operationality of the system.[60]

While Goux explains the possibility of transforming the logico-historical process into a new set of relations linked to a historicized nature, Baudrillard describes a domestication process where possibility only exists as an illusory last glimpse of "nature" as it disappears into the "reinforced operationality of the system," from which it will not reappear. What has emerged in this analysis is a sharp contrast between a developmental sense of history under the sign of utility (Goux), and a view of history based on fracture and dislocation leading to a domestication of human meaning and a consequent loss of community (Baudrillard).

Baudrillard's characterization of modernity — that it is ensnared in endless accumulation, and hence unable to reproduce itself symbolically in this context — is a confirmation of my original contention concerning a double disappearance. Not only is the natural world disappearing in a real sense, but the human impetus to conserve what is left of nature is also disappearing. Working from a frame of reference which first accounts for the transformative processes of modernity in a non-deterministic way, then describes this process as the disembeddedness of both economics from social institutions, and values and ideas from the relations of the things themselves, I have attempted to present the emergent and defining forms which have remade social relationships in the modern period.

This double disappearance has been expressed by Goux in terms of a culture of substitution, and by Baudrillard as the political economy of the sign. The generalized disembeddedness of this condition can be described as the universalization of economic processes (such as reification and commodity fetishism). These processes are no longer recognized as specifically economic, but instead refer to the pervasive displacement or sublimation of meaning and relations in modern society. It is this domesticated condition that is at the heart of modern society's inability to mount viable arguments for the conservation of nature.

Conclusion

Norman O. Brown describes the development of civilization as the triumph of dead matter over the living. The transformation of use-value into commodity, of commodity into money, of money into a general equivalent, of the general equivalent into generative capital, and of generative capital into a separation of value and sign from the relations of the things they represent, have charted an archaeology of the economic value form which confirms Brown's statement. Baudrillard's contention that modern society cannot reproduce itself symbolically because it is caught up in the processes of endless accumulation, can then serve as a definition for the crisis of modernity. The ultimate triumph of endless accumulation is that it ceases to be specifically economic as it implicates *all* expressions of value. I contend that it is this value form which is both promoting the destruction of the natural world, and also inhibiting — through its enigmas and mystifications — a form of analysis which would identify it as doing so. Once again, when I say that accumulation is no longer specifically economic, I am not saying that there are no specifically economic imperatives which continue to operate in society. There most certainly are. What I am saying is that the enigmas and mystifications which support these massive inversions are all but complete. It therefore becomes increasingly difficult to find a "frame of reference to which the market itself is referable."

Baudrillard's critique of the "giant reflexivity" of modern society sets the stage for extending social analysis to include the domain of nature. By undermining the modern economic assumptions based on production and signification which have informed human society's perspective in the modern era, and by viewing them as forms of human domestication, Baudrillard allows the sociality of nature to be considered as a basis for analyzing modern discourse on conserving natural communities. The social relationships which imbue nature with its significance can be seen as elaborations of Baudrillard's critique which rests on the "reversibility of accumulation and waste," and "cycles of giving and receiving."

If I begin with the assumption that the first human settlement did not suddenly and completely construct human reality but instead, that groups adapted past forms rooted in nature to their new surroundings, then it becomes possible to rewrite the insights of social theorists from a more nature-oriented paradigm. In their attempts to provide an alternate set of relations with which to analyse modern political economy, social theorists such as Marx developed theories about the mutual dependencies of use-values; Polanyi described forms of integration based on reciprocity and redistribution; Simmel spoke of the unbroken unity of the fruitful moment; Brown of expiation through sharing; and Goux of the immanent organizational potency of nature. When combined with Baudrillard's cycles of giving and receiving, these ideas show themselves to be buried references to a sociality that is also present in natural communities.

NOTES

1. Daniel Miller. 1987. *Material Culture and Mass Consumption*. London: Basil Blackwell, p.3.
2. Jean Baudrillard. 1975. *The Mirror of Production*. St. Louis: Telos Press, p. 121.
3. Miller (1987:3).
4. Miller (1987:3).
5. Miller (1987:4).
6. Arjun Appadurai. 1986. *The Social Life of Things*. Cambridge: Cambridge University Press. p. 5.
7. Benjamin Franklin. 1958, in *The Protestant Ethic and the Spirit of Capitalism* by Max Weber, New York: Scribners, p. 49.
8. Benjamin Nelson. 1969. *The Idea of Usury*. Chicago: University of Chicago Press.
9. Nelson (1969: xxv).
10. Karl Marx. 1959. *Capital*. Moscow: Foreign Language Publishing, p. 146.
11. Marx (1959:148).
12. Marx (1959:151).
13. Marx (1959:154).
14. Marx (1959:154).
15. Michael Taussig. 1980. *The Devil and Commodity Fetishism in South America*. Chapel Hill: University of North Carolina Press, p. 25.
16. Robert Heilbroner. 1985. *The Nature and Logic of Capitalism*. New York: Norton, p. 36.
17. Marx (1959:152).
18. Heilbroner (1985:37).
19. Taussig (1980:25).
20. Taussig (1980:33).
21. Taussig (1980:31).
22. Fredric Jameson. 1990. "Postmodernism and Consumer Society," in *Postmodernism and its Discontents*, edited by E. Ann Kaplan. New York: Verso, p. 10.
23. Jameson (1990:11).
24. Georg Simmel. 1990. The Philosophy of Money. New York: Routledge, p. 19.
25. Jean-Joseph Goux. 1990. *Symbolic Economies*. Ithaca: Cornell University Press, p.3.
26. Goux (1990:4).
27. Goux (1990:68).
28. Goux (1990:86).
29. Goux (1990:90).

30. Goux (1990:125).
31. Goux (1990:124).
32. Goux (1990:125).
33. Goux (1990:131).
34. Goux (1990:196).
35. Goux (1990:196).
36. Goux (1990:227).
37. Goux (1990:228).
38. Goux (1990:234).
39. Goux (1990:237).
40. Goux (1990:238).
41. Baudrillard (1975:20).
42. Baudrillard (1975:46-47).
43. Baudrillard (1975:22).
44. Mark Poster. 1975. "Introduction" to *The Mirror of Production* by Jean Baudrillard. St. Louis: Telos Press, p. 7.
45. Goux (1990:10).
46. Baudrillard (1975:114).
47. Baudrillard (1975:27).
48. Baudrillard (1975:29).
49. Baudrillard (1975:45).
50. Baudrillard (1975:48).
51. Karl Polanyi. 1968. "The Economy as Instituted Process," in *Primitive, Archaic, and Modern Economies*, edited by George Dalton. New York: Doubleday Anchor, p. 174.
52. Baudrillard (1975:49-50).
53. Baudrillard (1975:99).
54. Baudrillard (1975:102).
55. Baudrillard (1975:120).
56. Baudrillard (1975:127).
57. Baudrillard (1975:129-30).
58. Baudrillard (1975:131).
59. Baudrillard (1975:143).
60. Baudrillard (1975:146).

5
HUMAN IDENTITY AND THE NATURAL WORLD

> The material products of culture — furniture and cultivated plants, works of art and machinery, tools and books - in which natural material is developed into forms which could never have been realized by their own energies, are products of our own desires and emotions, the result of ideas that utilize the available possibilities of objects. It is exactly the same with regard to the culture that shapes people's relationships to one another and to themselves: language, morals, religion and the law. To the extent that these values are interpreted as cultural, we distinguish them from such levels of growth of their innate energies that they may accomplish, as it were, by themselves, and that are only the raw material for the process of civilization, like wood and metal, plants and electricity. By cultivating objects, that is by increasing their value beyond the performance of their constitution, we cultivate ourselves: it is the same value-increasing process developing out of us and returning back to us that moves external nature or our own nature.
>
> *Georg Simmel*[1]

The disruptiveness with which I have presented the expansion of the logic of general equivalent within modernity, contradicts the ascendant form related to the social construction of meaning and the cumulative approach to knowledge. So far I have traced this disruption in terms of the ways in which the emergent and defining form related to the expansion of capital, leads to a culture of substitution. To return to Raymond Williams'[2] representation of social change, new and emergent social realities — like the expansion of capital — transform the dominant paradigm. Within the transformation are residual forms of relations, which are ultimately lost to the emergent dominant discourse because they can no longer find expression in these new realities.

It is this residual cultural record of the loss of human identification with nature that I wish to take up. Until now, I have presented a generalized version of the expansion of capital and the ramifications of this expansion as it redefines relations within modernity. To describe this universalizing and abstracting process, I have had to engage in a form of discussion which tends toward the universal and the abstract. In order to locate this process more specifically, I will examine three specific points in the evolution of modernity which generated intense social change full of what Benjamin refers to as "moments of danger."[3]

In the context of this loss of human identification with the natural world during the modern period, I will examine the relationship between the emergent form of mercantile and agrarian capitalism in late sixteenth century England, and the consequent loss of human identification with the natural world as expressed in the Tragic literature of William Shakespeare. Secondly, I will look at the expansion of production and markets during the Industrial Revolution, and relate it to the loss of natural identity as expressed by John Keats' Romantic poetry. These preliminary discussions will lead to a third and more detailed discussion about the relationship between the globalisation of world economy and radical environmentalism, as current expressions of emergent and residual forms within modernity.

I have chosen to discuss these specific periods and writers because they represent — in the midst of rapid social change — a dialogue between competing world views. While certain sectors of society are regarding change in emergent and triumphant terms, others are expressing the grief of residual loss for the previous social formation. It is the recognition of this residual cultural record which can begin to counter the increasingly narrow conceptions of human identity and its relationship to nature, characteristic of many discussions on conservation.

Wild Convergence:
The Sociality of Natural Communities

As I have already stated, the central basis for the argument I am making rests on a social conception of nature. It is this social conception of nature, coupled with the insights of social theorists, which can begin to extricate us from the intellectual ruin of modernity. By using the word convergence in relation to sociality, it is my intention to make clear that — in the context of the arguments I am making — there is little difference between "nature" and the "social." If one of the defining aspects of modernity is separation and fracture by way of disembedding processes, then to speak of nature as something separate from human culture is to fall inside modern categories once again.

In broad terms, the modern period began by using terms such as "natural" and "unnatural" to describe human and nonhuman behaviour. The modern period is ending alongside the emergence of various forms of "development": developing countries, underdeveloped countries, the decade of development, sustainable development, and many more except, curiously, overdevelopment. So to speak of nature at all is to find oneself in transit at a way station. The nature that is out there is entered through the being experiencing it, not so much in oppositional terms but in a located seamlessness. As Simmel states, self and other are born in the same act, and that act ends the "fruitful moment." It is clear that if there ever was such a thing as the "fruitful moment," it ended long before the beginning of the modern period. What did begin to happen in the modern period is that — as Brown conveys very well — social forms changed from being re-integrative within nature, to being competitive and expansionary. The reified categories of capital which support competition and expansion thingify nature, turning the verb into a noun, as it were. When nature becomes a noun, things can be done to it, and we can then begin to talk about it. To reify capital into thinghood, is to reify nature into a noun. What this shift from a noun to a verb conveys is that the generative qualities of life which used to be "natural," has now been taken over by "development." The deadening process has replaced the enlivening one.

The de-animating relations described by Polanyi and Baudrillard in their discussion of capital and production become significant when nature is given social standing, along with its own generative qualities. For Baudrillard, it is communication rather than production which forms the basis of analysis "to which the market itself is referable." As Gregory Bateson states, communication in mammals "is primarily about the rules and the contingencies of relationship":

> What was extraordinary — the new great thing — in the evolution of human language was not the discovery of abstraction and generalization, but how to be specific about something other than relationship.[4]

Bateson's comment, "the general mammalian characteristic of [communication] being primarily about relationship,"[5] represents for me a socially-embedded conceptualization of nature. It is this emphasis on interrelationship which can generate a dynamic conception of human-nature relations.

My representation of nature as a social place relies most significantly on the work of naturalist John Livingston. In setting forth to explain his social conception of nature, Livingston lays out four categories (or "envelopes") of self which he sees as permeating the world of wild, undomesticated beings. As Livingston states, "the role of self in nonhuman beings is profoundly different from its role in human cultures — especially in the Western tradition."[6] The sense of "ecologic place" and "belonging" that exists in wild nature is in sharp contrast to the processes of placelessness related to the money form which intensified during the modern period. To quote Livingston:

> It is the wholeness of the wild animal that distinguishes it from the experientially deprived domesticate. It is the wholeness of the wild animal that makes ethical constructs unnecessary — indeed, probably unthinkable. Why create an abstract set of rules and guidelines when you are already doing all the right social things, and always have? Why seek replacement parts when you are complete? Rules and guidelines are for domesticates. Infantile, self-centred domesticates.[7]

It is in this sense that I understand Goux's statement concerning cultures of substitution: "It is only through replacement that value is created."[8] It is also this social sense of natural community which extends "the realm beyond economic value."

The "effectively lived" envelopes of identity that Livingston observes as existing in wild beings are as follows:

> * Individual self — infantile, primitive, placeless, ego-centred, competitive, acquisitive, unecological.
> * Group self — nuclear-family centred, cooperation limited to core group, competition suspended only within group, place limited to social, xenophobic, "territorial," unecological.
> * Community self — interspecies, cooperative, pro-

> found sense of communal, social and physical place, communal identity, participatory, ecological, wild.
> * Biospheric self— transcendent, whole, total connection, total belongingness.[9]

While these forms of natural identity are "simultaneously present and available in all wild beings at all times," Livingston argues that today, domesticated human identity is based almost entirely on the infantile and placeless individual self. It is this individual self that has adapted to, and been created by, the reified world of modern society. Furthermore, it has all but destroyed the other aspects of natural identity as the processes of market economy strengthened in the modern period. This resulted in a culture where

> . . . the infantile fixation on individual self which so perversely, it seems to me, is held in such high esteem in our culture. . . . Human groups, communities and populations are blessed to the extent that are aggregations of blessed individual souls, not as deserving entities in their own right.[10]

This individuated perspective is at the heart of modern economies and is reflected in liberal democracy's social and legal institutions. To quote Benjamin Barber:

> The problem of resorting to the law, any kind of law, in a modern constitutional liberal democracy is that the central character becomes the "legal person" and is always the most powerful entity. The standing of the family in such an arena will always be compromised by the fact that those who make up the family — child, wife husband — have stronger legal standing as individuals than as a family. How — in a system that atomizes us, privatizes us, and rewards us as legal persons — do we reinforce social entities that are based on kinship.[11]

This kind of individuality-based human society is in sharp contrast to what Livingston sees as a naturalist, when he looks at undisturbed natural communities:

> . . . naturalists perceive in the demeanor of a wild animal less a self-interested single entity, more a mani-

> festation of whole *being*. The naturalist sees a wild animal as one among uncountable ephemeral corporeal emergences, one minor miracle to remind us of the ineffable whole.[12]

A flock of shorebirds is not a composite of individuals, but a group-conscious super-organism. As Livingston states,

> it [the flock] is composed of discrete physical parts, to be sure, but its behaviour flows from one awareness, one consciousness, one self. In a psychological sense at least, the individual *is* the group.[13]

As such, this form of consciousness is not centred on the individual, but is "transcendent and *participatory*."

Alongside the existence of a self-as-group consciousness, there can be a self-as-community consciousness. This perspective poses a further challenge to the individually-based perspective on modern society:

> Perhaps the word "self" need no longer be used here. As we have seen, the role of the "other" has already diminished as we entered the envelope of the group consciousness. At the level of the community, "other" is bereft of whatever abstract meaning or utility it may have had. In the functioning multispecies community, all participants are subjects . . . . There need be no other; the community is a whole unto itself.[14]

Livingston's statement — "the community is a whole unto itself" — is the extension that I spoke of regarding Polanyi's and Baudrillard's "reciprocity of the group," or groups which contrast and can form the basis for, a critique of the "abstract meaning" of modernity. I believe that the statement, "in the functioning multispecies community, all participants are subjects," can form the basis for a social conception of nature which contradicts that of late modernity in which all participants are objects.

Depending on the nature of the species and the season, wild beings tend to identify with one of these envelopes of self, more than another. Migrating birds, for example, are not fixed to a specific place, but participate as a group consciousness that takes its cues from global occurrences, such as weather patterns and the amount of daylight. As well as positing a self-as-group consciousness and a self-as-community consciousness, Livingston believes that the only way to explain the "regional, continental and planetary promptings" to which migratory birds respond, is to suggest that there is a biospheric identity:

> Such a form of self-awareness would be transcendent — independent of the necessarily material context of individual, group, and community. But in moving beyond these, the biospheric self would not replace them. Rather it would subsume and hold them precious, as vital and integral domains, or types, of the greater whole.[15]

This fluid conception of the sociality of natural communities as being transcendent and participatory is the social understanding of nature which anchors my analysis. It is this sociality which is so manifestly absent from most discussions concerning the relationship between human society and the natural world. It is this perspective of embedded sociality — where "all participants are subjects" as opposed to commodities which relate through the "universal use of money as a mode of communication" — that can serve as the starting point for a social critique of contemporary discourse on managing the Earth .

As an expression of this loss of human identity, Livingston's paradigm of the complex relationships of wild beings, moves the discussion beyond the current impoverished categories, categories which confirm and replicate what amounts to the collective insanity of modern human society. In an attempt to upset epistemologies which do not acknowledge the social legacy of nature, I will argue that human social relationships have remained viable only because they have been underwritten by the sociality of natural communities (as identified by such naturalists as Livingston). It is the invisibility of the generative social aspects of nature which Goux identified as the "unthought absolute" of modern liberal-idealist visions of human-nature relations.

The Residual Cultural Record of Human-Nature Relations

> Ideas of nature, but these are the projected ideas of men. And I think nothing much can be done, nothing

> much can even be said, until we are able to see the causes of this alienation of nature, this separation of nature from human activity. . . .
>
> *Raymond Williams*[16]

Writing of the stories created by the peasants in South America as they try to respond to the incursion of capitalist modes of production, Taussig discusses the "intricate manifestations that are permeated with historical meaning and that register in the symbols of that history. . . ."[17] This "collective representation of a way of life losing its life" invokes a situation in which meaning and its expressions become discussible simultaneously as they are lost. Theorists such as Taussig argue that this loss does not exist in a use-value economy because, ". . . in a use-value economy the relations that persons enter into in their work appear to them as direct reciprocal, personal relations and not as activities controlled by the relationships of their products."[18] As an example of this kind of use-value culture, Taussig describes the society and beliefs of the Andean Indians on the slopes of Mount Kaata:

> To the Andean Indians residing on the slopes of Mount Kaata, the mountain is a human body. Their fields are used and their different products are exchanged in accord with the different functional parts of that body. The conception of nature and society as fused into the one organism is here most explicit. The land is understood in terms of the human body, and the human body is understood in terms of the culturally perceived configuration of the land. . . . The sacredness of the mountain is dependent on its wholeness: of nature, of the social group, of the person, and of all three together. . . . This is not the knowing of an exterior world by interior thought, as Cartesian epistemology and its legacy would have us understand. Instead it is "the omniscience to understand the secrets of the mountain body in terms of the corporeal body. Earth and humans no longer exist as dichotomies but rather as endless reflections of differently shaped mirrors." [Bastien 1978:43] To know is to be associated with everything around one and to enter into and be part of the land.[19]

The form of understanding which corresponds to this set of relations is, for Taussig, analogical reasoning:

> Analogical explanations involve an account of the unfamiliar in terms of the familiar, and the analogical mode of reasoning that is at issue here is inherently holistic and dependent on identifying things by their relationships to larger wholes.[20]

I contend that Taussig's description of the analogical reasoning in a use-value culture is not so much a reflection of the human social creation of reality, as it is a reflection of the immanence of natural sociality. To quote Taussig again:

> The analogical mode of reasoning is compelling in use-value economies because things are seen not as their self-constituents but as the embodiments of relational networks. Things interact because of meanings they carry - sensuous, interactive, animate meanings of transitiveness - and not because of meanings of physical force locked in the privatized cell of self-enclosed thinghood. . . . Thus the specific meaning of any of the terms within the total structure is dependent on the total set of relationships. This is to say that the significance of the individual terms is *not* a result of their meaning as isolates, disconnected from other isolates. Rather, they are relational terms that embody the meaning established by a set of relationships of which any term is a part. Things are relationships, and these relationships are ontological rather than logical.[21]

I think it is possible to link Taussig's statement that "things are relationships, and these relationships are ontological," with Livingston's statement that "all participants are subjects," in order to show them as statements of social embeddedness as well as to argue that Taussig's writing is about representing the "transitive" or the verb-oriented social legacy of nature, a legacy which continues to be represented in use-value culture.

What I have tried to show here is the strong connection between Livingston's natural being, and Taussig's descriptions of human-nature relations in earlier forms of social relations. Rather than seeing it strictly as an expression of use-value economies as Taussig maintains, I wish to upset the dominant view of reality by arguing that such concepts as analogical reason and similitude, relate to what Livingston has described as existing in wild nature. It is the anthropocentric perspective which assumes that all meaning is socially created by humans, and that humans brought nothing with them from their previous relationships with nature. It is this kind of displacement and inversion that has permeated so much of modern history. The invisibility of nature in this kind of construct dismisses the manifold ways in which human identity is rooted in natural being. It is also this kind of invisibility which has shown most current discussions about conserving nature as mirrors of impoverishment helping to destroy nature.

I recognize the ideological difficulties inherent in saying that what Livingston is representing is "nature," and that what others are describing as nature are really the amputated versions of an anthropocentric human identity. In our darker moments, every representation is fraught with pathology. But pathology is a construct too, and it need not dominate the social field. I reaffirm my contention that the recognition of the pervasive provisionalness of meaning in late modernity is a massive social failure that is particular to a specific culture. I also return to the statement that, whatever else the world may be, social theory has a lot to say about particular forms of exploitive relationships and the kinds of identities which go along with those relationships. As Williams stated in the opening quotation for this section, not much can be done until we understand the reasons for the separation of nature from human activity. Livingston's description of nature attempts to struggle with this separation, as well as providing ways of understanding relationships and identities which transgress the division between nature and culture.

Modern Divergence: The Split Between Nature and Culture

> In the first place it [economic transformation] had destroyed the two closely connected ideas of the community and the universe, and had replaced them by the two totally different concepts of the isolated individual and of infinite space.
>
> *Lucien Goldmann*[22]

Up until now, the processes of capital and modernity have been represented in generalized terms that do not reflect the idiosyncratic context of historical change. I believe this generalized approach has served the useful purpose of focusing attention on the emergent and generalizing form of capital which has transformed modernity in social terms. By contrast, attention to the historical moment is the best way to link the expansion of the emergent form of capital with the residual record of the loss of human identification with the natural world. This attention to the contingencies of historical change makes possible the opportunity of locating specific groups or classes which express these historical transformations in social terms. In periods of increased change and dislocation, there is a greater possibility for the self-conscious expression of this change emerging from the breakdown of consensus, which Benjamin describes as "moments of danger." This breakdown in consensus due to the increased pressures of historical change, has been represented at different times by way of civil wars or revolutions. I intend to examine the social transformations within modernity, by focusing on the contemporaneous expressions of emergent and residual forms in times of flux.

I will discuss three historically constituted "moments of danger": the emergence of English agrarian capitalism and its relationship to Shakespeare's *King Lear*; the expansion of the Industrial Revolution in England, and its representation in John Keats' poem "Hyperion"; and (in Chapter 6) the current movement towards the globalisation of world economy alongside the radical environmentalist commentaries on the relations between humans and nature. By discussing these "moments of danger," it is my contention that the cultural record of modernity shows representations of "ways of life losing their life." Specifically, I will show that representations of the older perspective are being threatened by a new perspective based on ambition and improvement. Whether it is in terms of eclipsing older strengths by younger strengths as in *King Lear*, or the defeat of the Titan gods by the Olympian gods in "Hyperion," or the overrunning of wilderness by resourcist exploiters, the idea that instrumentalism destabilizes the previous unimproved unity — a transformation presented by Simmel[23] in the opening quotation of this chapter as linking nature and human identity — is expressed in all these works.

What makes these expressions of "moments of danger" social transformations in which there is a residual record of loss, is the profound discordance between these residual manifestations and the new historical realities. Goldmann describes the Tragic perspective in this way: ". . . the complete and exact understanding of the new world created by rationalistic individualism. . . and at the same time, the complete refusal to accept this world."[24] In *Perspectives on Romanticism*, David Morse describes a similar refusal:

> What Romantic discourse foregrounds as a problem is the incommensurable: there is no common measure between . . . the poet's vision and the everyday, between the values of the past and the values of the present.[25]

Refusal and incommensurability run through all the discourses of radical environmentalism, and are expressed in Livingston's *Fallacy of Wildlife Conservation* when he states, "I rejoice in wildlife and I despair, in equal measure."[26] Together, there is a simultaneous recognition of identification with a world and the incommensurability of that identification with changing historical realities. It is the representation of this fracture in the "moment of danger" that is residually recorded as it is lost.

I believe that it is this residual history of modernity which can best inform the current environmental crisis. As expressions of what Williams defines as "experiences, meanings, and values, which cannot be verified or cannot be expressed in terms of the dominant culture" because they are expressions of a "previous social formation," the reclamation of what has been lost is an essential step in defining the character of the current environmental crisis. The recognition of this loss of sociality also makes it clear that humans and nature are in this together, and whatever happens to one also happens to the other. In fact, humans and nature are descriptive categories which are directly linked to this impoverishment. We may need new terms of analysis to extricate discussion from the current crisis of modernity.

Economic Conditions in Early Modern England

> The replacement of corporate entities by individuals as the constituent units of society, the separation of the state and civil society, the autonomization of the

> 'economy' — all these factors associated with the evolution of English capitalism conduced to the atomization of the social world into discrete and separate theoretical spheres.
>
> *Ellen Meiksins Wood*[27]

The emergence of capitalism is a particularly volatile ideological territory which, from time to time, is rendered asunder by the tectonic shifts of the present. Whatever else it may have meant in earlier periods to different sectors of society, the emergence and expansion of capitalism can currently be seen as the intensification of the exploitation of the natural world. With regard to this intensification of exploitation of nature, there is no historically constituted agency which develops with the means of production as is the case with certain social classes, and only an ever-increasing invisibility, both in terms of our understanding of it within the categories of capital, as well as its conversion into consumer products.

Works that can provide insights into this perspective on capitalism are those which focus on the dislocations brought on by the expansion of markets, rather than those which see the growth of capitalism as being inevitable. Since the central contention of this work is that there have been profound transformations in social relations within modernity, I do not accept that capitalism emerged from feudalism in some natural and self-evident process, nor do I accept a characterization of human identity as something which is predisposed to competition and efficiency, and which finally found full expression once the fetters of superstition and immobility were swept away by markets. To see history as a single trajectory is to read the assumptions of capital and modernity into the past in such a way that it collapses the discursive space which I have created here. As Wood states, "capitalism is no longer a presupposition. . . it emerged as an unintended consequence of relations between non-capitalist classes."[28]

As noted earlier in the chapter, I have chosen to focus on specific "moments of danger" in which there is societal instability and increased sense of contestations about social reality. Within these contestations, emergent and residual forms enter into a sharply drawn dialogue which will redefine the dominant perspective. In seventeenth century England, it is the residual expression found in Tragic literature and in the writings of radical sects such as the Diggers, Ranters, Seekers, and Levelers, which represent social groups that are losing their voice. It is this loss of voice which is temporarily in dialogue with other Protestant sectarians who align themselves with the emergent set of economic relations, and are in the process of taking a more dominant place in society. By focusing on the residual loss of sociality within these "moments of danger," the emphasis here is on dislocation and the loss of embedded sociality, rather than innovation and increased production.

Like all crises that tear at the fabric of a social order, the period that led up to the English Civil War in the mid-seventeenth century was a profoundly ambivalent time. Christopher Hill characterizes the upheaval in England this way:

> There were, to oversimplify, two revolutions in mid-seventeenth-century England. The one which succeeded established the sacred rights of property, . . . gave political power to the propertied, and removed all impediments to the triumph of the ideology of the men of property — the Protestant ethic. There was, however, another revolution which never happened, though from time to time it threatened. This might have established communal property, a far wider democracy in political and legal institutions, might have disestablished the state church and rejected the Protestant ethic.[29]

What Hill characterizes as the two revolutions which threatened England's for a time, represent for me the emergent and residual expressions of contrasting aspirations in certain sectors of the society, one of which had a defining role to play in the transformation of society, while the other receded into anachronism and only found temporary social expression before it was lost to society. It is this temporary expression of loss of social relations which I will examine in *King Lear*.

Much has been written about the specific circumstances for the emergence of capitalism in Early Modern England. The recent debate between social historians centres around Robert Brenner's claims that

> . . . the original breakthrough in Europe to a system of more or less self-sustaining growth was dependent upon a two-sided development of class relations:

> first, the breakdown of systems of lordly surplus extraction by means of extra-economic compulsion; second, the undermining of peasant possession or the aborting of any trend towards full peasant ownership of land. The consequence of this two-sided development was the rise of a novel social property system, above all on the land, in which for the first time, the organizers of production and the direct producers (sometimes the same persons) found it both necessary and possible to reproduce themselves through a course of economic action which was, on a system wide scale, favourable to the continuing development of the productive forces.[30]

As opposed to the standard Whig versions of the emergence of capitalism which assumes the gravitation of history toward the constellation of capitalism, as well as earlier Marxist analysis which focuses on the significance of the emergence of the bourgeoisie in social transformation, Brenner locates the crux of social change in the expansion of agrarian tenant farming which at the time, was entirely dependent on the market. Lacking both the aristocratic rights for extracting surplus from peasants, and the security of land ownership that was afforded to many peasants, this group of tenant farmers were wholly dependent on the local markets to sustain themselves, which in turn resulted in an increased incentive to expand agrarian production. It was also the generation of this agrarian production which allowed for the growth of towns and the increased division of labour since less people had to spend time on the land to produce their own food. As Wood states:

> The early modern English commercial system was thus distinctive in several respects: it depended on a highly developed domestic market and not just on foreign trade, on what might be called intensive rather than extensive expansion, the nexus of increased productivity and a growing market for consumer goods created by the agricultural revolution.[31]

It was these kinds of changes which caused economic relations to be increasingly driven "by the imperatives of productivity and competition within a single market."

These kinds of market changes led to widespread economic instability. To quote Terry Eagleton's *William Shakespeare*:

> Prices in England rose five times between 1530 and 1640; by the 1550s, after a century or more of relative price stability, agricultural prices were 95 per cent above those of the 1530s, and two-and-a-half times higher than prices forty years before. industrial prices in the 1550s rose by some 70 per cent over their level in the 1530s. The 1590s witnessed some enormous price fluctuations, and the average level of agricultural prices rose by one-third. The era of Shakespeare's dramatic production was rife with intense anxiety over a radically destabilized economy, and ridden with speculations about its causes. The debasement of the coinage, in the sense of the deterioration of the precious metal content of money, was widely held to be responsible for the mid-sixteenth-century financial crisis. Since food and fuel prices rose more sharply than those of other commodities, the result was a 'savage depression of the living standards of the lower half of the population.'[32]

Hill characterizes these transformations in terms of the creation of growing "masterless" groups within the eroding hierarchies of the medieval world, a world in which more and more people were put off their land in an effort to increase the production of food as well as other goods:

> The essence of the feudal society was the bond of loyalty and dependence between lord and man. . . . The assumptions were those of a relatively static agricultural society, with local loyalties and local controls. . . and by the sixteenth century society was becoming relatively mobile — masterless men were no longer outlaws but existed in alarming numbers — 13,000, mostly in the North, a government inquiry calculated in 1569; 30,000 in London alone, it was guessed more wildly in 1602.[33]

London's population expanded eight-fold between 1500 and 1650 as the victims of this dislocation moved into cities.

In the countryside, large groups had been cut loose because of these new economic realities. To quote Hill again, this masterless group included

> . . . rejects of a society in economic transformation, whose population was expanding rapidly. The necessity to economize led lords to cut down their households; the quest for profit led to eviction of some tenants from their holdings, the buying out of others. The fluctuations of the early capitalist cloth market brought wealth to a fortunate few, ruin to many. The inefficient and the unlucky went to the roads.[34]

The rural equivalent to the urban poor who had lost their place, took refuge in the remaining forest commons which had not yet been put to use by the extension of private property:

> . . . these were victims of the rapid expansion of England's population in the sixteenth century; sometimes the victims, sometimes the beneficiaries of the rise of new or the growth of old industries. . . . They specialized in robbing those who ground the faces of the poor, enclosers of commons, usurers foreclosing on land, 'builders of iron mills that grub up the forests with timber trees for shipping,' cheating shopkeepers and vintners; but not rent-racked farmers, needy market folks, labourers, carriers or women.[35]

Among those reacting to the changes occurring in Early Modern England was the Leveler, Gerrard Winstanley. Although he rejects the pieties of feudalism and kingship, Winstanley's communist views are not so much those of an "enemy of the existing social order,"[36] as David Wooten argues, but rather, a response to social transformations which were disembedding relations at the time. Winstanley states:

> A traditional Christian, who 'thinks God is in the heavens above the skies . . . worships his own imagination, which is the devil'. . . .[By contrast] God or reason can be known in man or nature, and man is more important than abstractions.[37]

For Winstanley, the societal changes that were occurring expressed themselves in terms of the conflict between community and selfish ambition: "Covetousness swells most against community, calling community a thief that takes other men's rights and property from them."[38] For Winstanley on the other hand:

> True freedom lies where a man receives his nourishment and preservation, that is in the use of the Earth .... True human dignity would be possible only when communal ownership was established, and buying and selling of land and labour ceased.[39]

Although Winstanley was one of the most radical proselytizers, egalitarian views were very common in the works of the Levelers, Ranters, Diggers, Seekers, and the Quakers of the time. These views of changing social conditions in Early Modern Europe are very different from those set forth by the "possessive individualism" usually associated with the Protestant ethic. Like Shakespeare's Tragic vision, these sectarians raise a voice of protest and refusal, the faint echo of which is heard presently in opposition to the promoters of free trade and global economies. It is also the voice of protest spoken in the name of human identification with the natural world.

The social transformation against which the radical sectarians protested is expressed in a general sense in the growing acceptance for the practice of collecting usury from fellow Christians. Calvin arrived at the notion of the permissibility of usury through an appeal to individual conscience, the equity of the Golden Rule, and the requirements of public utility: these are modern concepts not unlike the later calls for liberty, equality, and fraternity. To quote Benjamin Nelson's *The Idea of Usury*:

> In the pamphlets of the late sixteenth century, we are no longer in the atmosphere of the confessional; we are in the counting house. Henceforth, the controversy over the legitimacy of interest tends more and more to be carried on in eminently secular contexts, free from the otherworldly associations which had clung to the issue in the past.[40]

This presentation of the contentiousness of the issue of usury parallels Brown's psychoanalytic model which describes the shift from archaic consciousness to a modern consciousness:

> In the archaic consciousness the sense of indebtedness exists together with the illusion that the debt is payable; the gods exist to make the debt payable. Hence the archaic economy is embedded in religion,

> limited by the religious framework, and mitigated by the consolations of religion — above all, removal of indebtedness and guilt. The modern consciousness represents an increased sense of guilt, more specifically a breakthrough from the unconscious of the truth that the burden of guilt is unpayable. . . .
>
> Thus the increased sense of guilt brings about the emancipation of the economic process from divine controls and divine ends.[41]

It is this transition which brings forth the grief and loss expressed in the Tragic vision. What appears to many as the seemingly inevitable triumph of capitalism in the modern world was engendered in the anguish of dislocation, which in turn points to the underpinnings of a disappearing world. With regard to late twentieth century conservation discourses, such an approach sheds light on the loss of embedded social relations currently impoverishing much of the discussion about the relationship between human society and the natural world.

The Outskirts of Meaning: Nature and Tragedy in *King Lear*

It is *King Lear*'s[42] division into "older strengths" and "younger strengths" that groups the characters in the play, as well as forming the basis upon which the corresponding visions of "nature" reflect the conflicts that were occurring in early seventeenth century England. But what makes the play a Tragedy, and the reasons why it evokes the crisis constituting the loss of an interconnected world, is Lear's movement as King in full command of his throne, to Lear — crazed and full of night on the heath — consulting that "Noble Theban" Tom a Bedlam as to the cause of thunder.

The play opens with Lear displacing himself as he forfeits the responsibilities of sovereign in order that "we" may "unburdened crawl toward death." In so doing however, Lear proclaims that "Only we shall retain/The name, and all th' additions to a King. The sway,/Revenue, execution of the rest,/Beloved sons, be yours;" (I,i,137-140). By this act, Lear is "clutching at an empty signifier."[43] This elaborate inquiry into the relation between nature and culture begins with Lear retaining the name of King, but leaving open for contestation what that name means in a world that begins as a ritual to insure "that future strife/may be prevented now," and ends in a confusion of meaning and a disintegration of authority among a large group of younger strengths.

Within the full trajectory of the play, Lear's search for a signification which appropriately describes the title of King, leads him from the throne to Poor Tom a Bedlam ("the thing itself; unaccommodated man"). All this in an attempt to find some authentic meaning in nature, much in the same way that Macbeth's search leads him to the natural world of the three witches.

In this dissembling process, Lear becomes both King and Fool, proclaimer and commentator, an embodiment of the same and the other. It is a world of crisis and breakdown. Although he is not describing Lear, Foucault's comments (in *The Order of Things*) about the general process of change during this period are especially apt:

> Once similitudes and signs are sundered from each other, two experiences can be established and two characters appear face to face. The madman . . . is the man who is alienated in analogy. He is the disordered player of the Same and the Other. He takes things for what they are not, and people for one another, he cuts his friends and recognized complete strangers. . . he is constantly under the impression that he is deciphering signs: for him, the crown makes the King.[44]

If, as I have argued earlier, analogical reasoning is part of earlier forms of social relations and natural communities, this alienation in analogy is evidence that certain forms of understanding have failed to provide a viable description of the world. Instead of conveying similitude, they convey dysfunction.

By willfully creating a vacuum under his feet, Lear is reflecting a society moving into a state of flux. By displacing himself, he has fractured the relationship between semiotics and hermeneutics, and opened up for negotiation, competing epistemological perspectives. By moving through society from the throne to that place on the heath in a storm at the fringe of society where nature and culture contend, and where humans must outface with their nakedness, the inclemency's of the sky, the play reflects a moment when it is possible to confront all the social and religious constructs of a society from the perspective of "nature." As Eagleton remarks about Macbeth, "he becomes a floating signifier in ceaseless, doomed pursuit of an anchoring signified."[45]

The troubled relationship between humans and nature in the late twentieth century reverberates in these earlier contestations between aspiration and meaning, place and worth, obligation and ambition, competition and community, sign and meaning, and finally, between natural and unnatural. The contentiousness of value is set forth in the first scene by the all or nothing love of Lear's daughters. Both Goneril and Regan compete to outdo each other in terms of showing Lear the extent of their love for him. In response to this squandering, Cordelia can only "Love, and/be silent." and, in reply to her father's call to outdo her sisters, can only reply with "Nothing." Lear responds by saying "Nothing will come of nothing. Speak again." Signifying nothing is one of the preoccupations of the Tragedies, and in the case of Lear, the "name" of King as a semiotic sign confronts the hermeneutic "nothing" which now corresponds to it, and which forms a major focus of the play. "Nothing" finds "the thing itself" as Lear attempts to give the name of King a meaning.

On either side of Lear's movement from throne to heath, are the younger and older strengths. The contrasting perspectives of these two groups are embodied in the words of Gloucester and his bastard son, Edmund. The interconnected world of feudal hierarchy is expressed by Gloucester's commentary on the play's ongoing events and betrayals:

> *Glou*. These late eclipses in the sun and moon portend no good to us. Though the wisdom of nature can reason it thus and thus, yet nature finds itself scourged by the sequent effects. Love cools, friendship falls off, brothers divide. In cities mutinies; in countries, discord; in palaces, treason; and the bond cracked 'twixt son and father. This villain of mine comes under the prediction; there's son against father; the King falls from bias of nature; there's father against child. We have seen the best of our time. Machinations, hollowness, treachery, and all ruinous disorders, follow us unquietly to our graves (I,ii,102-113).

Gloucester's world of the "opposeless wills" of the gods is threatened by the treachery of his bastard son Edmund, who "was got betwixt unlawful sheets" and "came somewhat saucily into the world." Edmund presents a very different view of nature:

> *Edm.* Thou, Nature, art my goddess; to thy law
> My services are bound. Wherefore should I
> Stand in the plague of custom and permit
> The curiosity of nations to deprive me,
> For that I am some twelve or fourteen moonshines
> Lag of a brother? Why bastard? wherefore base?
> When my dimensions are as well compact,
> My mind as generous, and my shape as true,
> As honest madam's issue? Why brand they us
> With base? with baseness? bastardy? base, base?
> Who, in the lusty stealth of nature, take
> More composition and fierce quality
> Than doth, within a dull, stale, tired bed,
> Go to the creating a whole tribe of fops
> Got 'tween asleep and wake? Well then,
> Legitimate Edgar, I must have your land.
> Our father's love is to the bastard Edmund
> As to the legitimate, if this letter speed
> And my invention thrive, Edmund the base
> Shall top the legitimate; I grow; I prosper.
> Now, gods, stand up for bastards! (I,ii,1-22).

Upon overhearing Gloucester's laments cited above, Edmund scoffs:

> *Edm.* This is the excellent foppery of the world, that, when we are sick in fortune, often the surfeits of our own behavior, we make guilty of our disasters the sun, the moon, and the stars; as if we were villains on necessity; fools by heavenly compulsion; knaves, thieves, and treachers by spherical predominance; drunkards, liars, and adulterers by an enforced obedience of planetary influence; and all that we are evil in, by a divine thrusting on (I,ii,116-124).

What we see in the two contrasting perspectives as presented by Gloucester and Edmund is the fracture of the individual from nature, or "community and universe" as described earlier. Up until this point, humans are embedded in a culture describable as an unbroken web which includes the natural world. Hence, characters like Gloucester find it impossible to define humanity separate from the workings of the universe in this context. Edmund's speech expresses a kind of "arrogance of humanism" which the current destructions of the natural world can only interpret as a significant step in the loss of human identification with the natural world. Edmund's nature has hints of the Social Darwinism that has come to be known as an expression of modern capital.

Aside from his dedication to duty and service, Gloucester's only act of self will is demonstrated as he tries to kill himself by leaping from a "chalky borne" when his grief becomes too great. In contrast to Gloucester's life of service, the bastard son Edmund contends, in a letter he composed in the name of his brother Edgar, that "I begin to find an idle and/fond bondage in the oppression of aged tyranny" (I,ii,51-52). He is eventually slain for treason by a nameless opposer. As he dies, he affirms that "Yet Edmund was beloved./The one the other poisoned for my sake,/and after slew herself" (V,iii,241-243). He is referring to Lear's daughters and heirs, Goneril and Regan, who were competing for his love.

As expressed in the lives of Gloucester and Edmund, *King Lear* is neither a reaffirmation of feudal hierarchy nor a heralding of emergent individualism. As an expression of the Tragic vision, it is the embodiment of the transitional relationship between two world views, and a reflection of the relationship between nature and culture which arises out of such a transformation. This grappling for a meaning to signify the name of King, is the basis of Lear's "climbing sorrow" that "wrench'd my frame of nature/From the fix'd place," This sense of the Tragic vision is therefore the non-sense of the cracked molds of Lear's rantings. In light of the twentieth century's destruction of the natural world, it is the sense of this anguish and loss of human identity which permeates now.

Lear begins the play as very much the King on his throne who proclaims with the royal "we":

> *Lear*. For, by the sacred radiance of the sun,
> The mysteries of Hecate, and the night;
> By all the operations of the orbs,
> From whom we do exist, and cease to be (I,i,116-119).

By retaining only the name of King, these proclamations highlight the empty oaths of a human being who can no longer call upon the gods (who have retreated from the world) for authority. The Fool comments on the King's condition:

> *Lear*. Dost thou call me a fool, boy?
> Fool. All thy other titles thou hast given away; that
> thou wast born with (I,iv,146-148).

As Lear's world begins to disintegrate, the collapse of the external hierarchy becomes mirrored in Lear's internal conception of emotion:

> *Lear*. O, how this mother swells up
> toward my heart!
> Hysterica passio, down, thou climbing sorrow,
> Thy elements below! (II,iv,55-57).

Crucial to the disintegration of Lear's world is Goneril and Regan's circumscription of the small retinue of soldiers which he has retained along with the name of King, as a token of his position. When he loses them, he moves beyond the social realm and willingly casts himself into nature. In doing so, he comments on the relation between human society and the natural world:

> *Reg*. What need one?
>
> *Lear*. O, reason not the need! Our basest beggars
> Are in the poorest thing superfluous.
> Allow not nature more than nature needs,
> Man's life is cheap as beast's. Thou art a lady:
> If only to go warm were gorgeous,
> Why, nature needs not what thou gorgeous wearest,
> Which scarcely keeps thee warm. But, for true need—
> You heavens, give me that patience, patience I need!
> You see me here, you gods, a poor old man,
> As full of grief as age; wretched in both. . . .
> I will have such revenges on you both
> That all the world shall — I will do such things —
> What they are yet, I know not; but they shall be
> The terrors of the Earth! (II,iv,297-316).

There is an interesting contrast between the utility of Regan's perspective and Lear's refusal to succumb only to "need," which I believe echoes the position of the rentier aristocracy who were increasingly undermined by the culture of improvement. For Lear, "the terrors of the Earth" refer to the total collapse of the hierarchical world he once knew. After experiencing the storm on the heath, this is exactly what he calls for, albeit now as a crazed King with only a Fool to serve him:

> *Lear*. Blow, winds, and crack your cheeks! rage! blow!
> You cataracts and hurricanoes, spout

> Till you have drenched our steeples, drowned the cocks!
> You sulph'rous and thought-executing fires,
> Vaunt-couriers of oak-cleaving thunderbolts,
> Singe my white head! And thou, all-shaking thunder,
> Strike flat the thick rotundity o' the world,
> Crack Nature's molds, all germans spill at once,
> That make ungrateful man! (III,ii,1-9).

"Ungrateful" can be read as Winstanley's "covetousness" or Brown's suppression of guilt, and reflects the loss of obligation and community. Having called for an end to his world by a cataclysmic apocalypse in which nature's elements overwhelm the human world, Lear recognizes that he now exists in a very different relation to the forces governing that world:

> *Lear.* I tax not you, you elements, with unkindness.
> I never gave you kingdom, called you children,
> You owe me no subscription. Then let fall
> Your horrible pleasure. Here I stand your slave,
> A poor, infirm, weak, and despised old man
> (III,ii,16-20).

This departure from the social world and the entry into the natural world, is expressed when Lear seeks shelter in a sheep's hovel:

> *Lear.* Where is this straw, my fellow?
> The art of our necessities is strange,
> And can make vile things precious. Come, your hovel
> (III,ii,72-74).

This insight leads Lear to contemplate the lowest of the low in his spent kingdom:

> *Lear.* Poor naked wretches, wheresoe'er you are,
> That bide the pelting of this pitiless storm,
> How shall your houseless heads and unfed sides,
> Your looped and windowed raggedness, defend you
> From seasons such as these? O, I have ta'en
> Too little care of this! Take physic, pomp;
> Expose thyself to feel what wretches feel,
> That thou may'st shake the superflux to them
> And show the heavens more just (III,iv,35-43).

As compared to the writings of radicals such as the Leveler Winstanley quoted earlier, it is clear that a very different view of the relationship between humans and nature is being presented here than is usually associated with either the medieval or modern perspective. It is a cataclysmic and egalitarian vision taking its cue from the forces of natural elements, which Lear now thinks he has found embodied in the person of "the thing itself," Poor Tom:

> *Lear.* Thou wert better in thy grave than to answer
> with thy uncovered body this extremity of the skies. Is
> man no more than this? Consider him well. Thou owest
> the worm no silk, the beast no hide, the sheep no wool,
> the cat no perfume. Ha! Here's three on's are sophisticated!
> Thou art the thing itself; unaccommodated man
> is no more but such a poor, bare, forked animal as thou
> art. Off, off, lendings! Come unbutton here (III,iv,106-113).

Lear sheds his clothes in an attempt to expunge the last vestige of his social identity and enter the authentic world of nature. It is a world where Tom

> . . . eats the swimming frog, the toad,
> the tadpole, the wall-newt and the water; that in the fury
> of his heart, when the foul fiend rages, eats cow-dung for
> sallets, swallows the old rat and the ditch dog, drinks the
> green mantle of the standing pond; (III,iv,131-135).

But at the same time as Lear attempts to leave culture for nature, he refers to Tom as the "learned Theban," "Athenian," and "philosopher;" highlighting the fact that Lear continues to use cultural categories. Also, Poor Tom "the thing itself" is a fiction and disguise created by Gloucester's son Edgar.

When Lear carries the recently hanged Cordelia onstage in his arms, he is also carrying a world that has died: "She's dead as Earth" (V,ii,264). "Thou'lt come no more,/Never, never, never, never, never!" (V,iii,309-310). Here is the nothing that comes from nothing. Here is the meaning that can attach itself to the name of King. This is the significance of the transformation of social realities in Early Modern England that is expressed in Tragic literature. Aliveness has left the Earth. What remains is a limited and relative world that is administered by a proxy. In spite of Edgar's attempt to behold the changes in social reality, he cannot fail to draw a pale moral that will dress nature as culture:

> *Edgar*. The weight of this sad time we must obey,
> Speak what we feel, not what we ought to say.
> The oldest hath borne most; we that are young
> Shall never know so much, nor live so long (V,iii,389-392).

The limited world is affirmed and accepted, almost with relief, as the new King Albany orders the bodies of Lear and Cordelia off the stage: "Bear them from hence. Our present business/is general woe" (V,ii,320-321).

In the same way that Brown describes the disappearance of the devil from modern perspectives, as the invisibility of the price paid by humans to live in the modern world, the retreat of god from the world — the central motif of Tragedy — is the disappearance of the "immanent organizational potency" that weaves together a world in which there is no separation between human culture and the natural world. Tragedy can only exist as long as the gods who are called upon do not appear. It is the withdrawal of this dimension from human existence which is expressed in Tragedy. The gods are what link community and universe in a seamless web, even though they embody at the same time what has been lost to human sociality by displacing that sociality upwards, when once it had been diffused throughout the world as a set of embedded relations. Relationships are ontological. All participants are subjects.

The retreat of the gods from human affairs is transformed downwards into a commiseration with the embodied representation of "the thing itself." As I have stated earlier, "the thing itself" is not an essence but a set of embedded relations. The play's linking of the retreat of the gods with the loss of embedded relations which links nature and culture in the things themselves, makes it an incisive chronicle of the social transformations occurring in early seventeenth century England which have led to the deadening of the Earth.

The Manufactured World

> What is essential is that at the beginning of the nineteenth century a new arrangement of knowledge was constituted, which accommodated simultaneously

> the historicity of economics (in relation to the forms of production), the finitude of human existence (in relation to scarcity and labour), and the fulfillment of the end of History. . . . Finitude, with its truth, is posited in time; and time is therefore finite.
>
> *Michel Foucault*[46]

Whereas the fracture and loss in Tragedy is that of a King who can no longer sway the gods, it is from the lonely platform of the individual that seeks momentary reverie which is characteristic of Romantic loss. But in either case, the significant fact is what Morse[47] referred to as the incommensurability of the conflict between two competing views of the world. Something was irretrievably lost, and I maintain that this loss was part of the humanness which identified itself with the natural world, and which found the new economic realities to be irredeemably in conflict with that identity. Both Tragic literature and Romantic poetry, as well as current expressions in radical environmentalism, can be understood as part of a process that recognizes the disembedding of social relations, and as such, are residual cultural records within modernity of the loss of human identification with the natural world.

In his book on the Industrial Revolution entitled *Continuity, Chance and Change*, E. A. Wrigley states:

> By an astounding irony of modern historiography, the Industrial Revolution, whose intrinsic interest and importance should make it the most exciting topic of study among all the 'big' issues of the history of the development of the modern world, has become a dull subject that slips into focus and out again, uncomfortably peripheral to the vision of many historians.[48]

The Industrial Revolution had become "a convenient label" to attach to the changes occurring at the turn of the nineteenth century in England. The significance of the Industrial Revolution becomes increasingly obvious as modern society faces

> . . . a world that has gained an awesome momentum of growth but may lose any semblance of stability.

> Such has been the legacy of the Industrial Revolution. . . . We cannot choose but to be the inheritors of the Industrial Revolution; we can choose to know our inheritance better than we do.[49]

Polanyi offers a description of the changes brought about by the Industrial Revolution in stark terms:

> We submit that an avalanche of social dislocation, surpassing by far that of the enclosure period, came down upon England; that this catastrophe was the accompaniment of a vast movement of economic improvement; that an entirely new institutional mechanism was starting to act on Western society; that its dangers, which cut to the quick when they first appeared, were never really overcome; and that the history of nineteenth century civilization consisted largely in attempts to protect society against the ravages of such a mechanism. The Industrial Revolution was merely the beginning of a revolution as extreme and radical as ever inflamed the minds of sectarians, but the new creed was utterly materialistic and believed that all human problems could be resolved given an unlimited amount of material commodities.[50]

This ability to create an unlimited amount of material commodities went beyond what had been the real limits of an agricultural economy, as seen by the Physiocrats, and so generated goods without overstepping the limits of "the land." What Wrigley describes is the new open-endedness of a mineral-based energy economy:

> Thus the typical industries of the new regime produced iron, pottery, bricks, glass, and inorganic chemicals, or secondary products made from such materials, above all an immense profusion of machines, tools and consumer products fashioned out of iron and steel. The expansion of such industries could continue to any scale without causing significant pressure on the land, whereas the major industries of an organic economy, textiles, leather and construction, for example, could only grow if more

> wool, hides or wood were produced which in turn implied the commitment of larger and larger acreages to such ends, and entailed fiercer and fiercer competition for a factor of production whose supply could not be increased.[51]

It was precisely this kind of increasingly capital intensive industries which required a constant supply of raw material to make them profitable. This supply could only be insured by opening up wider and wider areas to the unregulated market. The requirements of large investment brought more and more people into the market economy and subjected an increasing range of activities and goods to the standards of measuring labour until:

> All transactions are turned into money transactions, and these in turn require that a medium of exchange be introduced into every articulation of industrial life. All incomes must derive from the sale of something or other, and whatever the actual source of a person's income, it must be regarded as resulting from sale.[52]

This process can be defined as the disembedding of social relations within market dynamics which control access to the means of production. This has profound ramifications for social relationships where "it means no less than the running of society as an adjunct to the market."[53]

In attempting to represent this social transformation in *The Making of the English Working Class*, E. P. Thompson chooses to focus on the residual cultural record of the blind alleys, the lost causes, and the losers of the history of the period, much in the same way that Christopher Hill discusses the English Civil War in the context of the more radical revolution envisioned by the Diggers and Ranters. It is these naysayers and footdraggers who offer a perspective which can extricate discussion from the autocracy of triumphal capitalism.

For Thompson, "the crucial experience of the Industrial Revolution was felt in terms of changes in the nature and intensity of exploitation"[54] in which there was "the reduction of the man to the status of an instrument."[55] Thompson characterizes the intensification of this exploitation in this way:

> The classic exploitive relationship of the Industrial Revolution is depersonalized, in the sense that no lin-

> gering obligations of mutuality — of paternalism or deference, or of the interests of 'the Trade' — are admitted. There is no whisper of the 'just' price, or of a wage justified in relation to social or moral sanctions, as opposed to the operation of free market forces. Antagonism is accepted as intrinsic to the relations of production. Managerial or supervisory functions demand the repression of all attributes except those which further the expropriation of the maximum surplus value from labour.[56]

It is these kinds of changes which hardened the separation between nature and human culture, a separation which Romantic literature attempted to represent.

Society as Graveyard

> Romanticism, then, and this is the best definition I can give it, is spilt religion.
>
> *T. E. Hulme*[57]

To quote Raymond Williams, "there have been few generations of creative writers more deeply interested and more involved in study and criticism of the society of their day"[58] than the Romantics. In fact, the contextual reality of the Romantic outlook was deeply rooted in changes in economic and social relations during the period:

> It is in fact in this same period in which the market and the idea of specialist production received increasing emphasis there grew up, also, a system of thinking about the arts of which the most important elements are, first, an emphasis on the special nature of art-activity as a means to 'imaginative truth,' and, second, an emphasis on the artist as a special kind of person. . . . Yet. . . the response is not merely a professional one. It is also (and this has been of the greatest subsequent importance) an emphasis on the embodiment in art of certain human values, capacities, energies, which the development of society towards an

> industrial civilization was felt to be threatening or even destroying.[59]

Young's *Conjectures on Original Composition* (1759) embodies what was considered authentic in contrast to what is manufactured and artificial:

> An Original may be said to be of a vegetable nature; it rises spontaneously from the vital route of genius; it grows, it is not made; Imitations are often a sort of manufacture, wrought up by those mechanics, art and labour, out of pre-existent materials not their own.[60]

It is this kind of inauthenticity that Simmel refers to in the opening quotation of this chapter when he states, "by cultivating objects. . . we cultivate ourselves."[61] What Simmel is making clear is that, by way of the Industrial Revolution, humans are manufacturing themselves. Human life is perceived to be becoming inauthentic. Society's behaviour in general, had taken on an artificial connotation because of the disembedding affect of economic relations. Grace is no longer something arbitrarily granted by god, but a state achieved by the soul as it "works" its way free of "the values of the present." Genius becomes the vital route back to nature. Individual uniqueness replaces the sense of universal reality because there now appears a standardizing of social activity based on commodification. More than anything else, Romanticism was focusing on "a mode of human experience and activity which the progress of society seemed increasingly to deny,"[62] and was constructed in response to "the aggressive individualism and the primary economic relationships which the new society embodied."[63]

Williams' essay, "Base and Superstructure in Marxist Cultural Theory," in which he outlines emergent and residual sectors of a dominant culture, states that "literature appears by no means only in the emergent sector, which is always, in fact, quite rare. A great deal of writing is of a residual kind. . ."[64] In the same way that Goldmann[65] portrays Tragedy in terms of loss, and a refusal to accept new economic and philosophical realities, Romanticism embodies a similar kind of recognition that the new realities of the emerging Industrial Revolution bring about a sense of loss and as a result, instills a "rebellion" of the human spirit which attempts to affirm and protect a set of social relationships which it recognizes as being under threat.

This sense of transformation in social realities, much like the contrast between the older strengths and the younger strengths in *King Lear,* is illustrated clearly in John Keat's poem "Hyperion"[66] where the gods of the Titans are dethroned by the new Olympian gods. The poem begins:

> Deep in the shady sadness of a vale
> Far sunken from the healthy breath of morn,
> Far from the fiery noon, and eve's one star,
> Sat gray-haired Saturn, quiet as a stone,
> Still as the silence round about his lair; (I:1-5).

Saturn, who has become his own gravestone, dies because a vale has passed between himself and nature, not just as the event of death, but also in terms of understanding and relatedness. The stream has become "voiceless," "still deadened more by reason of his fallen divinity spreading a shade" (I:8-9). This fallen divinity, this place where culture and nature meet, is portrayed by "His old right hand [which] lay nerveless, listless, dead,/Unsceptred;" (I:18-19), while his bowed head "listen[s] to the Earth,/His ancient mother, for some comfort yet" (I:20-21).

Thea, goddess of the infant world and wife of Hyperion, describes Saturn's condition in this way:

> For heaven is parted from thee, and the Earth
> Knows thee not, thus afflicted for a God;
> And ocean too, with all its solemn noise,
> Has from thy sceptre passed; and all the air
> Is emptied of thine hoary majesty (I:55-59).

In response to Thea's entreaties, old Saturn "lifted up/His faded eyes, and saw his kingdom gone,/And all the gloom and sorrow of the place. . ." (I:89-91). He then relates the great sadness of his condition:

> I am gone
> Away from my own bosom: I have left
> My strong identity, my real self,
> Somewhere between the throne, and where I sit
> Here on this spot of Earth. (I:112-116)

This quotation could serve as the theme of this work. It is also a commentary on the whole modern progressive human project.

Rather than reaffirming kingship, this statement can be said to focus on a placedness within an interconnected identity represented by the "throne." Like Tragedy, the throne seems to be lost first, and during this state of loss, Saturn finds himself on a "spot of Earth," unconnected with any other spot of Earth. The sense I get from the statement, "I am gone/Away from my own bosom," is that what had been a world of unbroken weaves, has now become fractured and arbitrary. The relationship between this perception and the events of the Industrial Revolution are unmistakable in their presentation of "the sorrow of the time" (I:301).

Hyperion, god of the sun, is one of the few Titan gods who has not yet succumbed to the "horrors new" (I:233). Coelus, Son of Mysteries, entreats him to go "To the Earth!/For there thou wilt find Saturn, and his woes" (I:345-346). Coelus describes the state of the present world as "Manifestations of that beauteous life/Diffused throughout the eternal space" (I:317-318). When this presentation of scattered meaning is contrasted with the intense "greenness" in the imagery of the rest of the poem, it is clear that the diffusing of manifestations "throughout eternal space," is a distinct part of nature's "ruin". As Coelus says of himself, "my life is but the life of wind and tides" (I:341). This kind of identification with nature's elements and forces is repeatedly referred to by the fallen Titan gods. Central to their fall is the diffusion of "beauteous life" throughout "eternal space."

The spectre of human mortality, or "human finitude" as Foucault called it, is central to this disembedding process. Mortality is presented as the reason Saturn has lost his identification with the natural world:

> . . . the supreme God
> At war with all the frailty of grief,
> Of rage, of fear, anxiety, revenge,
> Remorse, spleen, hope, but most of all despair.
> Against these plagues he strove in vain; for Fate
> Had poured a mortal oil upon his head,
> A disannointing poison: (II:92-98).

As well as echoing the "smell of mortality" which Lear experiences as a King full of grief and rage, this quotation presents the loss of nature in terms of a descent into human mortality. And, "as with us mortal men" (II:101), what is shared is that "other hearts are sick with the same bruise" (II:104). The reason for human mortality cannot be found in "the legends of the first days" (II:132), even though Saturn tries to "read it deep" (II:148). In this world which represents the "values of the past," there are no clues to human mortality:

> No, nowhere can unriddle, though I search,
> And pore on Nature's universal scroll (II:150-151).

This nature is the nature that the Titans saw as indivisible from the natural elements. As such, the Titans represent unimproved nature.

In response to Saturn's plea to not "unriddle" the cause of the Titan's plight, Oceanus replies that Saturn "must be content to stoop" (II:178) as there are successive stages in nature, and that Saturn must accept the fact that the Titans have been superseded. In the same way that the Titans, as representatives of "new and beauteous realms" (II:201) were an improvement on "chaos" and "darkness," now "on our heels a fresh perfection treads" (II:212). Oceanus states that "tis the eternal law/That first in beauty should be first in might" (II:228-229).

The only one to speak in response to Oceanus' resignation is Clymene. She does not challenge him, but instead complains "with hectic lips" that "I am here the simplest voice,/And all my knowledge is that joy is gone" (II:252-253). She describes how she heard music being played by Apollo and that "a living death was in each gush of sounds" (II:281). She decries to her fellow Titans: "O. . . had you felt/Those pains of mine" (II:296-297). As Apollo relates it, he dreamed of the goddess Mnemosyne, and when he awoke, he found a golden lyre at his side:

> Whose strings touched by thy fingers, all the vast
> unwearied ear of the whole universe
> Listened in pain and pleasure at the birth
> Of such new tuneful wonder (III:64-67).

What made Apollo the usurper of the Titans and what changed the music of the air, was improvement or the arts. Whereas the Titans were unimproved nature, Apollo, along with other Olympian gods, is "loveliness new born" (III:79), and the "fresh perfection" (III:79) as Oceanus describes them.

Apollo perceives the same kind of world which has already been conveyed by Saturn. He states:

> For me, dark, dark,
> And painful vile oblivion seals my eyes:
> I strive to search wherefore I am so sad,
> Until a melancholy numbs my limbs;
> . . . . Why should I
> Spurn the green turf as Hateful to my feet?
> . . . . Where is power?
> Whose hand, whose essence, what divinity
> Makes this alarum in the elements,
> While I here idle listen on the shores
> In fearless yet in aching ignorance?
> . . . . Tell me why thus I rave, about these groves!
> Mute thou remainst — mute! Yet I can read
> A wondrous lesson in thy silent face:
> Knowledge enormous made a god of me (III:86-113).

Whereas Saturn could not "unriddle" nature's scroll, Apollo is able to "read" the silent face of the god. Note that it is not within nature, but in reading the silent face of a god who has retreated from the world, that Apollo gains the knowledge which makes him immortal. At the same time as he reads the gods, he finds the "green turf" hateful to his feet, as if he wants to ascend from the natural world. The knowledge that Apollo gains is not that of the elements of an unimproved nature, but instead, that of

> Names, deeds, gray legends, dire events, rebellions,
> Majesties, sovereign voices, agonies,
> Creations and destroyings, all at once
> Pour into the wide hollows of my brain,
> and deify me, as if some blithe wine
> Or bright elixir peerless I had drunk,
> And so became immortal (III:114-120).

Immortality is not gained by being an element of nature, but by having knowledge. And it is not an elemental knowledge, but a chaotic and unstable knowledge of deeds and dire events — knowledge befitting a god of a new kind of world. In the Titans, there is the recognition of loss, and in the Olympians the Romantic strategies of survival. Perception is now in the "hollows of my brain." The world has gone inside, just as the god has retreated from the world. All this to nature's loss.

Although "Hyperion" expresses much of what is expressed in Tragic literature such as *King Lear* (including the fracture and loss associated with the "hidden god"), what is dramatically different is its idea of transcendence through knowledge. It is a deeply ambivalent process full of suffering, and tinged by mortality:

> Most like the struggle at the gate of death;
> Or liker still to one who should take leave
> Of pale immortal death, and with a pang
> As hot as death's is chill, with fierce convulse
> Die into life (III:126-130).

It is the unspoken striving for aliveness which is the key to remaining human. Romanticism is an attempt to retain a palpable human connection with the suffused aliveness of the natural world. It was, as Wordsworth calls it, "the rock defense of human nature"[67] against a world that is becoming increasingly antagonistic to it.

If, indeed, science cannot be expected to "save" nature, then it is important to resurrect the tradition of experiencing the loss of nature as the loss of human identity. It is the indivisibility of human identity and the aliveness of a living and diverse natural world, which can serve as the necessary correctives to the narrowing of discussions concerning the preservation of nature in the late twentieth century. Only that recognition will free the current discourse from what William Blake calls "the fiends of commerce."

Moments of Danger

What I have attempted to point out in my discussion of Tragedy and Romanticism, are particular "moments of danger" in what has been, in Chapter 3 and 4, a general and monolithic representation of the expansion of the logic of general equivalents in the modern period. This discussion of Tragedy and Romanticism is also an attempt to locate current radical environmentalism within a frame of a "way of life losing its life," which focuses on the connection between human identity and nature, and their transformations within modernity.

An interesting avenue into the resurrection of this loss is to observe the ways in which the proselytizers of the intensification of capitalist discipline reconceptualize human identity so as to denigrate "the passions" as a failure of the uninitiated, and as a stumbling block to the increased efficiency of economic relations. Thompson conveys this transformation at the start of the nineteenth century in terms of views of children as innocent beings, to beings of "a corrupt nature and evil dispositions," as well as "unsubdued passion" who need to be brought into the intensification of the "work-discipline of industrialism."[68]

This transformation is also very clear in the federalist-antifederalist debate in the eighteenth century with regard to the American constitution. Bertell Ollman states that the federalists supported the "constitution as a political extension of the capitalist mode of production."[69] Central to this project was the attempt to depoliticize the people by excluding them as much as possible from direct involvement in the political process. This was necessitated by the federalist belief that the people are given — in the view of Madison and Hamilton — to "irregular passions" which makes the political process unpredictable. It is this apparent dichotomy between passion and efficiency which can be reframed when we consider Sheldon Wolin's discussion about the antifederalist opposition to federal notions of centralization. Wolin states that for the antifederalists,

> . . . it was natural and desirable for their government to interfere in the economy. The economy was not a sacred object, but a set of relationships that might have to be amended when the good of the members required it.[70]

This connection between resistance to the disembedding categories of capital, and the denigration of those aspirations as "passions," reveals the way certain embedded aspects of identity are transformed, and sometimes demonized, by the discourse of capital.

It is this same denigration that is evident in Martin Lewis'[71] negative characterization of the emotionality of radical environmentalists as they resist capitalist discipline. By setting radical environmentalism within a tradition of "a way of life losing its life," it is my goal to question the disembedding frame that Lewis uses to analyse those passions. Like the "fiends of commerce" that have come before him, Lewis is a proselytizer of capital engaged in "decoupling" human society from nature.

Nature Rendered Unconscious

> [T]he question is not whether wilderness has a tomorrow but whether Homo Sapiens have a future without wild nature.
>
> *Max Oelschlaeger*[72]

The appearance of "the death of the subject" in postmodern critical theory, and the simultaneous discussion of "nature for its own sake" in radical environmentalism, provide an evocative basis for examining the present relation between human identity and the natural world. I regard these phrases as recognitions which, when seen in the context of the emergent and residual forms that move into and out of the dominant paradigm of modernity, point to industrial society's difficulty in mounting a viable argument for the preservation of a living and diverse natural world, and for a viable human identity. This is the third "moment of danger" and will be discussed at greater length in the following chapter.

The view that humans are nature rendered conscious, can be seen as a valorization of many of the destructive aspects of the modern project, or it can be regarded as a constructive contribution to contemporary discussion about the human relation to the natural world. Max Oelschlaeger states that "the idea of wilderness appears to undergrid a new paradigm for understanding humankind as embodying natural processes grown self-conscious."[73] But this perspective becomes very problematic when the viability of consciousness of self has been called into question (as represented in the idea of "the death of the subject"). As Fredric Jameson states in his essay "Postmodernism and Consumer Society":

> This new component [in postmodern discourse] is what is generally called the 'death of the subject'. . . today, from any number of distinct perspectives, the social theorists, the psychoanalysts, even the linguists, not to speak of those of us who work in the area of culture and cultural and formal change, are all exploring the notion that the kind of individualism and personal identity [as established in Modernity] is a thing of the past. . . .[74]

In his book *Postmodernism, or the Cultural Logic of Late Capitalism,* Jameson refers to cultural production in the context of the "death of the subject":

> Cultural production is thereby driven back inside a mental space which is no longer that of the old monadic subject but rather that of some degraded collective "objective spirit": it can no longer gaze

> directly on some putative real world, at some reconstruction of a past history which was once itself a present; rather, as in Plato's cave, it must trace our mental images of the past upon its confining walls. If there is any realism left here, it is a "realism" that is meant to derive from the shock of grasping that confinement and of slowly becoming aware of a new and original historical situation in which we are condemned to seek history by way of our own pop images and simulacra of that history, which itself remains forever out of reach.[75]

This scenario of human perception does not lend itself to any viable relationship between human and nonhuman nature, nor as a conceptualization of nature rendered conscious. For Jameson, the last connection between humans and "a phantasmatic relationship with some organic pre-capitalist peasant landscape and village society," was conveyed by Heidegger:

> Today, however, it may be possible to think all this in a different way, at a moment of radical eclipse of Nature itself: Heidegger's "field path" is, after all, irredeemably and irrevocably destroyed by late capitalism, by the green revolution, by neocolonialism and the megalopolis. . . . The other of our society is in that sense no longer Nature at all, as it was in precapitalist societies, but something else which we must identify.
>
> I am anxious that this other thing not overhastily be grasped as technology per se. . . . Yet technology, may well serve as adequate shorthand to designate that enormous properly human and anti-natural power of dead human labour stored up in our machinery — an alienated power, what Sartre calls the counterfinality of the practico-inert, which turns back on and against us in unrecognizable forms and seems to constitute the massive dystopian horizon of our collective as well as individual praxis.
>
> Technological development is however. . . the result of the development of capital rather than some ultimately determining instance in its own right.[76]

It is not worthwhile to try and state, in a Millinarian fashion, who was the last to express a social connection between humans and nature, but I would argue that beyond Heidegger, radical environmentalism is the last collective expression of a way of life losing its life that is connected to a wild nature.

An important aspect of this technological development which has an impact on authentic experience is the electronic media. Heidegger's "field path" becomes a very different kind of field of experience in the context of pervasive media. As Jay Rosen points out in his article, "Playing the Primary Chords":

> . . . the listener's or viewer's brain is an indispensable component of the total communication system. In other words, the audience isn't really "out there" at all. The goal can't be to reach the audience because there is no territory to reach across. Our minds are no longer separate cognitive spaces. They are each part of the electronic commons — a shared space, privately held but publicly traveled.[77]

As well as seeing the "death of the subject" from the perspective of cultural theory and in terms of electronic media, it is also possible to see how Livingston's[78] characterization of the domesticator-domesticate relationship which pervades urban industrial society, can also lead to a representation of modern identity in terms of total otherhood. In fact, what Livingston sets forth as the social qualities of the domesticator-domesticate relationship is the "one-way dominance" mirror which first created the privileged position of humans within its cultural ascendancy. The "unilateral dependence" of the domesticate, as portrayed by Livingston, can be read as the loss of agency of the human subject, whose social relations have lost their viability and it connects this loss of viability with the collapse of the human relation with the natural world.

The other phrase with which I began this discussion — "nature for its own sake" — is a recognition of the collapse of the modern human relation with the natural world. Only if humans are excluded from nature can nature be preserved. Any human interest is a destructive interest when it comes to the natural world. Modern industrial society is so caught up in the resourcist mode that a perspective that is interested in the preservation of a living and diverse natural world, has no recourse but to exclude those humans from their place in that world. Evernden's discussion in *The Natural Alien*, about human-nature relationships as part of a global niche in which there are only "humans and natural resources,"[79] leads to a similar conclusion pointing to the incommensurability of the inclusion of modern society in any vision for a viable natural world.

What "nature for its own sake" recognizes as a call for nature preservation, is that modern society exists in a world driven by "consumption for its own sake." What the simultaneous appearance of the "death of the subject" and "nature for its own sake" point to, is that a viable human identity, as with a diverse natural world, cannot be sustained in the current context of Goux's "technological society" or Baudrillard's "political economy of the sign," and that when these forces of production and consumption fracture the final links in the human relationship with nature, human meaning dies altogether. All subjects have become objects.

In a discussion of the life of drug addicts in New York City entitled "The Possessed", Luc Sante has this to say about human identity and drug addiction:

> The potent illusion that drugs provide is called upon when the more commonplace illusions fail, and especially when life appears as nothing more than the conduit between birth and death. Drugs populate the empty landscape, supply the missing heaven, extend the movie into the third dimension.[80]

I find this a very evocative presentation of both the breakdown of modern identity and the assumptions about the relationship between humans and nature, which is implicit in the statement concerning the reasons for that breakdown.

The interesting reversal here is the assumption that under normal conditions social customs or "commonplace illusions" provide the basis for making human life livable, and masks what would otherwise be a meaningless and empty void. It is the idea that humans grant meaning to the world through the creation of culture — naming the animals as it were — that has been the basis of the growing anthropocentrism of modern society. But, as this culture itself is swallowed up by the disembedded processes of capital, this bequeathing of meaning collapses.

Modern society has come to the end of a mistaken reversal. It is nature and natural identity which underpins much of the meaning for humans, and it has been the survival of our natural identities within culture, which has allowed for human meaning to continue to be viable. The "empty landscape" is not one in which social creation of meaning has failed, but one in which cultures of substitution have wrung the last vestiges of identification with nature from human relations.

NOTES

1. Georg Simmel. 1990. *The Philosophy of Money*. New York: Routledge, pp. 446-47.
2. Raymond Williams. 1980. *Problems in Materialism and Culture*. New York: Verso, p. 38.
3. Walter Benjamin. 1973. *Illuminations*. London: Collins, p. 257.
4. Gregory Bateson. 1972. *Steps to an Ecology of Mind*. New York: Ballantine, p. 367.
5. Bateson (1972:367).
6. John Livingston. 1994. *Rogue Primate: An Exploration of Human Domestication*. Toronto: Key Porter Books, p. 103.
7. Livingston (1994:103).
8. Jean-Joseph Goux. 1990. *Symbolic Economies*. Ithaca: Cornell University Press, p. 10.
9. John Livingston. 1994. This is unpublished compilation of one of Livingston's main arguments in *Rogue Primate.*
10. Livingston (1994:104).
11. Benjamin Barber. 1991. "Enough About Rights. What About Responsibilities?" *Harper's Magazine*. February, p. 49.
12. Livingston (1994:104).
13. Livingston (1994:107).
14. Livingston (1994:111).
15. Livingston (1994:117).
16. Williams (1980:82).
17. Michael Taussig. 1980. *The Devil and Commodity Fetishism in South America*. Chapel Hill: University of North Carolina Press, pp. 17-18.
18. Taussig (1980:129).
19. Taussig (1980:157).
20. Taussig (1980:135).
21. Taussig (1980:36-37).
22. Lucien Goldmann. 1964. *The Hidden God*. New York: Norton, p. 27.
23. Simmel (1990:446-47).
24. Goldmann (1964:33).
25. David Morse. 1981. *Perspectives on Romanticism*. London: Macmillan, p. 257.
26. John Livingston. 1981. *The Fallacy of Wildlife Conservation*. Toronto: McClelland and Stewart, p. 101.
27. Ellen Meiksins Wood. 1991. *The Pristine Culture of Capitalism*. New York: Verso, p.91.
28. Wood (1991:10).
29. Christopher Hill. 1974. *The World Turned Upside Down*. Harmondsworth: Penguin, p. 15.

30. Robert Brenner. 1985. *The Brenner Debate*. New York: Cambridge University Press, p. 214.
31. Wood (1991:99).
32. Terry Eagleton. 1986. *Shakespeare*. Oxford: Basil Blackwell, p.103.
33. Hill (1974:39).
34. Hill (1974:40).
35. Hill (1974:43).
36. David Wooten. 1986. *Divine Right and Democracy*. Harmondsworth: Penguin, p. 60.
37. Quoted in Hill (1974:141&142).
38. Quoted in Wooten (1986:326).
39. Quoted in Hill (1974:134).
40. Benjamin Nelson. 1969. *The Idea of Usury*. Chicago: University of Chicago Press, p. 83.
41. Norman O. Brown. 1985. *Life Against Death: The Psychoanalytical Meaning of History*. Middletown: Wesleyan University Press, pp. 271-72.
42. William Shakespeare. 1957. *King Lear*. Toronto: Washington Square Press.
43. Eagleton (1986:77).
44. Michel Foucault. 1970. *The Order of Things*. New York: Vintage, p. 49.
45. Eagleton (1986:3).
46. Foucault (1970:262).
47. Morse (1981:257).
48. E.A. Wrigley. 1988. *Continuity, Chance, and Change*. New York: Cambridge University Press, p. 2.
49. Wrigley (1988:6).
50. Karl Polanyi. 1957. *The Great Transformation*. Boston: Beacon Press, p. 40.
51. Wrigley (1988:5).
52. Polanyi (1957:41).
53. Polanyi (1957:57).
54. E. P. Thompson. 1980. *The Making of the English Working Class*. London: Penguin, p. 218.
55. Thompson (1980:222).
56. Thompson (1980:222).
57. Quoted in M. H. Abrams. 1971. *Natural Supernaturalism*. New York: Norton, p. 118.
58. Raymond Williams. 1982. *Culture and Society*. London: Hogarth, p. 30.
59. Williams (1982:36).
60. Quoted in Williams (1982:37).
61. Simmel (1990:447).
62. Williams (1982:30).
63. Williams (1982:42).

64. Williams (1980:44).
65. Goldmann (1964:33).
66. John Keats. 1973. "Hyperion," in *Romantic Poetry and Prose*, edited by Harold Bloom and Lional Trilling. New York: Oxford University Press, pp. 543-56.
67. Quoted in Williams (1982:41).
68. Thompson (1980:441).
69. Bertell Ollman. 1990. "Introduction," to *The United States Constitution* edited by Bertell Ollman and J. Birnbaum. New York: New York University Press, p. 5.
70. Sheldon Wolin. 1990. "The People's Two Bodies: The Declaration and the Constitution," in Ollman and Birnbaum (1990:132-133).
71. Martin Lewis. 1992. *Green Delusions: An Environmentalist Critique of Radical Environmentalism*. Durham: Duke University Press.
72. Max Oelschlaeger. 1991. *The Idea of Wilderness*. New Haven: Yale University Press, p. 327.
73. Oelschlaeger (1991:320).
74. Fredric Jameson. 1990. "Postmodernism and Consumer Society," in *Postmodernism and its Discontents*, edited by E. Ann Kaplan. New York: Verso, p. 17.
75. Fredric Jameson. 1991. *Postmodernism, or the Cultural Logic of Late Capitalism*. Durham: Duke University Press, p. 25.
76. Jameson (1991:34).
77. Jay Rosen. 1992. Playing the Primary Chords," in *Harper's Magazine*, March, p. 23.
78. Livingston (1994:100).
79. Neil Evernden. 1985. *The Natural Alien*. Toronto: University of Toronto Press, p. 110.
80. Luc Sante. 1992. "The Possessed," in *New York Review*, November 12, p. 23.

6
Prisoners of Value: The Current Environmental Debate

> Recognizing the human world as one of utility reveals an important truth: that this is a social world, in which nature appears as a humanized nature, i.e. as the object and material base for industry. Nature is the laboratory and raw material base for procuring, and man's relationship with it resembles that of a conqueror's relationship, a creator to his material. . . . Reducing the relationship between man and nature to that of a producer and his raw material would infinitely impoverish human life.
>
> *Karel Kosik*[1]

It is the goal of this chapter to apply the analysis of modernity I have presented so far to current calls for conservation of the natural world. For the purposes of the discussion here, I will focus on Martin Lewis' book *Green Delusions: An Environmentalist Critique of Radical Environmentalism.*[2] Lewis' work is particularly apt because it both mounts an argument for conservation based on sustainability, and also criticizes "eco-radicals" for not doing the same. By criticizing eco-radicals and at the same time making a case for moderate environmentalism, Lewis' work brings forth discussions about the frames of reference in which the relationship between human society and the natural world are understood, and therefore helps define what is meant by "environmental crisis." I will argue that Lewis' proposal to "decouple" human society from nature in order to save nature, is part of a process which I have described in terms of disembeddedness, a process which has led to the destruction of nature. Whereas I have characterized this process as central to the undermining of the modern human relationship with nature, Lewis advocates it as a strategy for conserving nature.

By examining works such as Lewis' from the perspective of the creation of cultures of substitution, it is not my goal to reject outright all of the conservation initiatives that these works propose. Instead, what I intend to do is present what I consider to be serious obstacles that are confronting industrial society as it attempts to understand the nature of the environmental crisis. I argue that the centrality of money in the modern period has not only caused many of the overt problems related to the devastation of nature, but it has also created the categories in which these issues are understood.

The ways in which modern economic relations have influenced these conservation discussions express themselves in terms of two apparent polarities: in the framing of discussions related to conservation and development; and in the link between conceptions of human identity and conceptions of nature. It is through a discussion of these polarities that I will show the extent to which the assumptions that underwrite conservation initiatives are implicated in economic realities which are destroying nature.

In making problematic this call for the integration of the perspectives of conservation and development, I will contextualize the perspectives of conservation and development as they have evolved within the history of modern society, and as part of what Robert Heilbroner in *The Nature and Logic of Capitalism,*[3] calls the "social form" of modern society. As discussed in Chapters 3 and 4, the transformations in the money form were identified as an emergent and defining form which has had a profound influence in shaping modern society. Conversely, the loss of human identification with the natural world (as discussed in Chapter 5) is presented as a residual form within modern society which has recorded this loss of identification occurring as a result of economic changes. A clarification of what conservation and development mean in the context of this frame of reference related to modernity, can shed light on disagreements and conflicts within the environmental movement about frames of reference. I contend that perspectives based on sustainability can best be understood as remaining within the emergent form of economic transformation, and that the call for the integration of conservation and development is for the most part pre-empted because this call remains trapped within the processes which have led to cultures of substitution. Conservation in this context refers to wise uses of resources and remains an instrumental process in development, if albeit, over a longer time-frame.

Conversely, I define the conservation of nature in terms of the recognition of commonality between human identity and nature, which has become a receding and residual social form in the modern period. It is in this context that the disagreements (discussed by Lewis) between moderate and radical environmentalists can best be understood, with moderates defining themselves in terms of emergent globalisation and ecosystem thinking, and radicals recognizing the social loss that has occurred between humans and nature in the context of increasing global economic realities. At its most extreme, it is the incompatibility between the residual perspective of radical environmentalism and the emergent perspective of modern economics which finds expression, not so much in terms of contrasting strategies toward environmental issues, but in terms of the way forms of sociality are understood from different social locations at a given time in history.

A consideration of conservation as representing this residual form, and development as representing the emergent form, points up the improbability of their being integrated. Instead, the appearance of conservation and development in this dialectic form is a record of a "moment of danger" which, within the forces of globalisation, human identification with nature — as expressed by "eco-radicals," to use Lewis' term — is temporarily experienced as loss in the context of new economic realities. This loss becomes invisible to those identities made in the new reality's image, and the temporary intensity of radical environmentalism dissipates as a particular relationship between human identity and nature disappears. By examining the relationship between conservation and development in the context of residual and emergent forms, the call for their integration is placed both in a historical context and is seen as an expression of social location within modern society (as in the North-South dialogue).

The other aspect of these works which I wish to focus on is the relationship between human identity and nature, as well as other issues related to their viability. Much of the contentiousness of the environmental debate revolves around defining the qualities of human identity and nature. For some, humans are forms of capital and resources, while nature is a large farm. For others, humans are an expression of four billion years of evolution within an indivisible nature which has only recently been torn asunder by modern society. Whatever basis is used to set the frame for discussions on conservation, there is a close link between who we are as humans and the kind of nature we think about conserving.

In the previous chapter, I discussed this relationship in terms of the failure of viable conceptions of human identity related to the "death of the subject," and arguments for conservation related to "nature for its own sake," a view which seeks to remove industrial humans from nature because they are so destructive. This view is a radical cancellation of a modern human identity as an expression of nature rendered consciousness. Once again, I contend that the qualities of human identity and the corresponding qualities attributed to nature, are central to an understanding of the kinds of humans and the kinds of nature that are to be conserved.

In arguing that economic forces related to the money form have led to what has been called the crisis of modernity, or that concepts such as sustainability offer little hope for conservation of the natural world, I am certainly not saying anything new. My contribution to the discussion on conserving the natural world, is the bringing together of these two fields of perspective in order to provide a particular frame of reference in which these discussions can be understood.

Economic Development as an Emergent Form

As noted earlier, this book is not about strategizing functional knowledge in the name of management; it is about a recognition of dysfunctionality leading to an analysis of sets of assumptions. When the environmental crisis is framed in terms of the integration of conservation and development, an analysis of assumptions is pre-empted and the focus is shifted to strategizing. What I mean by this is that from the perspective of this book, what needs to be examined in the current environmental crisis is the ways in which industrial society is destroying nature. This has nothing to do with integration or conservation. I contend that what has to be addressed is the massive inappropriateness of industrial society and the assumptions that support this inappropriateness. To discuss this reality in terms of integrating conservation and development is to accept — to use an inappropriate metaphor — the fox in the hen house. It is not so much the accommodation involved — a charge usually leveled at moderate environmentalists — as it is the pre-empting of our understanding of the assumptions which lead to destructive practices against nature, assumptions which also pervade a wide range of modern activities, including conservation initiatives, and end up amounting to little more than a struggle for resources carried on by other means.

The view of conservation as sustainability based on an integration of conservation and development is represented by a wide range of individuals and groups:

Maurice Strong, Secretary-General of United Nations Earth Summit in Brazil in June, 1992:

> The only way to economic viability and sustainability in the future is going to be a marriage of the environment and development.[4]

Ruth Grier, Ontario Minister of the Environment and Chair of "The Ontario Round Table on Environment and Economy:"

> The [Round Table] report puts us in the forefront of a world movement toward sustainable development. The health of the economy and the environment are indivisible.[5]

IUCN, UNEP, and WWF, authors of *World Conservation Strategy*:

> . . . bodies responsible for development and conservation should be analysed to assess the extent to which ecological considerations are incorporated into the development process.[6]

World Commission on Environment and Development, author of *Our Common Future*:

> Failures to manage the environment and to sustain development threaten to overwhelm all countries. Environment and development are not separate challenges: they are inexorably linked.[7]

MacNeill, Winsemius, and Yakushiji, authors of *Beyond Interdependence*:

> . . . the world has now moved beyond economic interdependence to ecological interdependence — and even beyond that to an intermeshing of the two.[8]

Robert Costanza, Herman Daly and Joy Bartholomew, editors of *Ecological Economics: The Science and Management of Sustainability*:

> To achieve sustainability, we must incorporate ecosystem goods and services into our economic

> accounting. The first step is to determine values for them comparable to those of economic goods and services. In determining values, we must also consider how much of our ecological life support system we can afford to lose.[9]

The World Conservation Union, United Nations Environment Programme and World Wide Fund for Nature, authors of *Caring For The Earth*:

> We need development that is both people-centred, concentrating on improving the human condition, and conservation-based, maintaining the variety and productivity of nature. We have to stop talking about conservation and development as if they were in opposition, and recognize that they are essential parts of one indispensable process.[10]

As one reads these quotations, it is almost possible to hear the doors of analysis of environmental issues creaking shut, and the universalizing of modern economic categories taking over. The call of eco-radicals such as Dave Foreman[11] collapses into invisibility when sustainability becomes the frame of reference.

It is my intention to link my presentation of cultures of substitutions with what is called development, in the context of the environmental crisis. By doing so, I intend to make problematic the inclusion of this paradigm of development in strategies for conservation (in ways that sustainability arguments do when they call for the integration of conservation and development). In *Dominating Knowledge*, Stephen Marglin describes the "sticking points" which inform a certain definition of development:

> . . . it is less the goals [of development] than the ***processes*** that matter, so we have little to gain from distinguishing between development and modernization. Of course, even with the focus on process, there still remains a diversity of views about what development and modernization mean. However we probably shall not go far wrong if we place the following at the core: on the economic side, industrialization and urbanization, as well as the technological transformation of agriculture; on the political side, rationaliza-

> tion of authority and the growth of a rationalizing bureaucracy; on the social side, the weakening of ascriptive ties and the rise of achievement as the basis for personal advancement; culturally, the 'disenchantment' of the world (to use Max Weber's terminology), the growth of science and secularization based on increasing literacy and numeracy.[12]

It is possible to link Marglin's emphasis on the processes of development, with Baudrillard's contention that substitution is "the very process"[13] of the political economy of the sign.

As Marglin states, ". . . the Western model of development, notwithstanding its considerable economic successes, has yet to produce an acceptable model for relationships between people or with nature."[14] Once again, it is possible to link Marglin's statement with Baudrillard's argument that "capitalism is only concerned with value and is unable to reproduce itself symbolically."[15] In fact, as Marglin points out, development is just as predatory to human cultural diversity as it is to the biological diversity of nature. Or to put it another way, the monolithic aspect of modern development has created a view of nature and culture which requires that they be described as diverse in order to contrast them with modern realities, when in fact it is actually the monolith which needs definition. The question "why preserve biological diversity?" is created by a monolithic narrowing of perspectives created by cultures of substitution, as if nature has to now justify why it should remain the way it has always been. This is what I mean when I say that most current discussions replicate the grid of general equivalents and inform a perspective which is able to formulate such a question as "why preserve biological diversity?" It certainly would not occur to the Yanomamo — a people who live in Central Brazil — to ask such a question. As an emergent and defining form that shaped modern society, economic development is problematic both in terms of its practice and the ways in which it defines the epistemological field in which issues such as environmental problems are discussed.

In his chapter entitled "Environment" in *The Development Dictionary*, Wolfgang Sachs describes the struggle between different frames of reference in this way:

> The eco-cratic discourse which is about to unfold in the 1990s starts from the conceptual marriage of

> 'environment' and 'development', finds its cognitive base in eco-system theory, and aims at new levels of administrative monitoring and control. Unwilling to reconsider the logic of competitive productivism which is at the root of the planet's ecological plight, it reduces ecology to a set of managerial strategies aiming at resource efficiency and risk management. It treats as a technical problem what in fact amounts to no less than a civilizational impasse — namely, that the level of productive performance already achieved turns out to be not viable in the North, let alone for the rest of the globe. With the rise of eco-cracy, however, the fundamental debate that is needed on issues of public morality — like how society should live, or what, how much and in what way it should produce and consume — falls into oblivion.[16]

It is this kind of framing of environmental issues in terms of conservation and development which Sachs sees as the intellectual "ruin" which inhibits an approach that promotes the conservation of a diverse natural world. Note that both Marglin and Sachs emphasize the strong link between the extrication of human identity from the categories of development and the extrication of nature from the resourcist paradigm.

Scientific American's *Managing Planet Earth* contains much of this kind a reformist approach which Sachs criticizes:

> World economies are depleting stocks of ecological capital faster than the stocks can be replenished. Yet economic growth can be reconciled with the integrity of the environment.[17]

The above quotation is the emboldened subtitle which introduces James MacNeill's article in *Managing Planet Earth* entitled "Strategies for Sustainable Economic Development." This approach is also reflected in Leonard's *Divesting Nature's Capital*:

> Our intention in this book is to examine how various natural resource and environmental constraints affect contemporary economic development strategies and prospects of the so-called developing nations in the world.[18]

Although the depletion of the natural world is seen as being of great significance for the well-being of present and future generations, it is also clear that the natural world — as described in terms of an economic balance sheet — is configured as an aspect of industrial capital. In works such as these, nature appears in either of two forms: as a sink for toxins or as a factory which produces an annual surplus for exploitation. This perspective assumes a conception of human identity which is also dependent on the economic balance sheet. This leads to the conclusion that this perspective is more interested in preserving industrial growth than preserving a diverse natural world. It is not even as if there was a choice of any other conception of nature or human identity than a resourcist one. It is the "emergent" approach which does not analyse industrial capital in a way that makes it problematic, but instead "strategizes" with the givens of that paradigm. To quote Colin Clark in *Mathematical Bioeconomics*:

> Recognizing the capital-theoretic nature of resource stocks is essential to a clear understanding of resource economics. From this viewpoint resource management becomes a special problem in capital theory, although it is an especially interesting and difficult problem.[19]

Within the development paradigm, nature is "a special problem in capital theory." This is clearly stated in Robert Repetto's *World Enough and Time: Successful Strategies for Resource Management*:

> Global resource issues will probably not be successfully addressed without the support and cooperation of business. Large international companies have the technology and management expertise for successful environmental and resource management. The business community also has obvious long-run interests in the sustained productivity of the resource base.[20]

Within the implicated paradigm of development, the categories of capital form the basis for all discussion, and nature appears only as a resource subject to the pressures of capital, which conservation attempts to extend into a longer time frame.

In Chapter 4, I discussed how the purely quantitative valuing of money had eventually altered the qualitative values of modern society. A logic of general equivalents — as expressed in economics, semiotics, metaphysics, and psychoanalysis — became the basis upon which the relationship between industrial society and nature is understood. In framing the discussion of modern society in a way that attempts to extricate discussion from those substitutions, the viability of human identity and the human relationship with nature are called into question. It is precisely these kinds of issues that arguments in favour of sustainability do not make problematic since they accept as given "the processes" of development.

Greenback Delusions

> But men labor under a mistake. The better part of the man is soon plowed into the soil for compost. By a seeming fate, commonly called necessity, they are employed. . . laying up treasures which moth and rust will corrupt and thieves break through and steal. It is a Fool's life, as they will find when they get to the end of it, if not before.
>
> *Henry Thoreau*[21]

In discussing Lewis' book, I will attempt to illustrate in more detail the contention that current discussions of environmental issues related to the exploitation of the natural world which focus on the integration of conservation and development and describe themselves in terms of sustainability, are still operating within the categories of industrial capital and are oriented toward a conception of human identity and nature as defined by the context of commodities and production. Works such as that of Lewis frame the relationship between human society and the natural world in terms of initiatives which operate in a market economy — such as the internalization of economic externalities and the possibilities of tradable pollution permits — as ways of "solving" environmental problems. It is these kind of strategies which attempt to adapt current economic practices so as to maintain, with the least disruption, a modern conception of human identity and lifestyle for those who already have it, and still provide a possibility for that segment of the world's population who aspire toward it. From this perspective, what is required is not the wholesale critique of market economy, but a reform of certain of its practices.

Lewis begins his work by stating that along with the threat posed by those who think the ecological crisis is a mirage which should not inhibit continued economic expansion, there is a threat posed to nature by radical environmentalists whose "ill-conceived doctrine" that "human society, as it is now constituted, is utterly unsustainable."[22] For Lewis, this leads radical environmentalists to contend that "society must be attacked 'at the root' and reinvented . . ." along lines based on "a more direct relationship between humanity and nature."[23] For Lewis, this reinvention poses a disruptive threat to both the stability of human society and the preservation of what is left of nature.

In contrast to this "eco-radical" conception of the ecological crisis, Lewis states:

> I would argue that the most ecological course for human society is, in fact, to divorce ourselves, and our economy, from the natural world. Our greatest contribution to the environment will come when we're farthest removed from it — in the laboratory, in the voting booth, and in the marketplace. To advocate this notion, which has been labeled "decoupling," is to acknowledge a profound division between humankind and the rest of nature — the very division that many Greens allege is at the root of the ecological crisis.[24]

Rather than rejecting the structures of modern political economy as predatory, Lewis states that it is by utilizing these market-based structures to institute such programs as effluent taxes and pollution permits that can be traded, as well as for converting toxic waste into valuable synthetics by such companies as Dupont, which will lessen human pressure on the natural world. These solutions will come ". . . if at all, from high-tech corporations — from firms operating in a social, economic, and technical milieu largely removed from the intricate webs of the natural world."[25] It is this orderly response to the ecological crisis that Lewis contrasts with the "shamanistic rituals" of deep ecologists who would dismantle modern society in an attempt to reclaim "a way of life losing its life."

It is not my goal here to decide whether Dupont or radical environmentalism offers the best solution to environmental problems, although I would argue that Lewis' contention that eco-radical solutions are unworkable and deluded, has far more to do with the massive inappropriateness of industrial society than with any failing of radical environmentalism. It is my intention to discuss Lewis' work in light of the analysis of the emergent and residual forms in modern society which have focused the concurrent processes of the social construction of culture related to the money form (of which the moderates are a part), and the residual loss of human identification with nature that has accompanied this process (as expressed by eco-radicals). I believe that the discord which Lewis presents as existing between "moderate environmentalists" and "green extremists," is a record of the passing of a "moment of danger" — to use Benjamin's term[26] — and that the experience of the loss of human identification with nature is central to this discord. I will therefore examine the way Lewis constructs the contrasting views of human identity and nature as stated by both the moderates and the extremists, to illustrate how economic processes and the loss of human identification with nature are reflected in the way these groups understand the environmental crisis.

Central to my presentation of modernity is the linking of the transformation identified as disembeddedness leading to a culture of substitution, with Lewis' self-conscious advocation of the decoupling of human society from nature in order to preserve nature. I contend that the disembedding processes that I have represented in exceedingly problematic terms for both modern human identity and nature, are understood by Lewis as the affirmative base upon which humans can now decouple human society from nature. His work is then a kind of fulfillment of the modern processes which I have been describing. The obvious contrast between his perspective and my own with regard to these processes, provides an opportunity to discuss the contradictory understandings of the environmental crisis represented by moderates and radicals.

In order to connect the current discussions of identity (in humans and nature) with previous moments of danger, I contend that — like the expression of loss in Tragic literature and Romantic poetry which occurred simultaneously with the acceleration of the emergent forms of capital in the beginnings of mercantile capitalism and the expansion of the Industrial Revolution — there is a social attachment to nature represented in the views of "eco-radicals" which is expressed as it is lost (in the context of rapid changes in technology and the increasing globalisation of economic realities). By regarding this social attachment to nature as a threat,

Lewis' perspective can be seen as one that has been remade in the context of these new economic realities. It is attached to these new realities, rather than to a departing residual vision of nature. Within the conflict expressed in *Green Delusions* between "moderate" commitment to working with existing institutions and corporations, and the contrasting "extremist" attachment to nature and estrangement from present social and economic realities, Lewis constructs in a dramatic fashion what I have described in terms of emergent and residual forms. What is to be wondered at here is that while there is a kind of synchronic acceptance of the conditions of late modernity by moderate environmentalists who carry on in their problem solving ways regarding environmental issues, eco-radicals are experiencing similar conditions in terms of incommensurability, and are throwing the cultural furniture of late modernity out the doors and windows.

It is not my goal in turn to show that Lewis is deluded and that his arguments are ill-founded. Instead, it is my goal to reveal the significance of the appearance of his argument in historical terms. If the history of modernity is a record of appearances which are temporarily discussible as they are lost, Lewis work stands as a record of the passing of the human attachment to natural community. By regarding the eco-radical linkage of human community with natural community as a threat to modern society, Lewis' work is a recognition and confirmation that human identity has been remade in the context of new realities related to globalisation which in a self-conscious way do not require nature. The momentum of his arguments is away from "seduction" and towards "logic." My analysis is not so much a partisan analysis of Lewis' main contentions — although I regard them as offering little hope for the conservation of the natural world — as it is an attempt to frame his presentation of the disagreements between moderate and radical environmentalists within a particular vision of the trajectory of modernity.

By contrast, Lewis outlines the controversies related to the environmental crisis in a straightforwardly partisan manner:

> I seek . . . to defend the broad-based environmental movement . . .[that remains] committed to reforming our economy and society through dogged work within normal political and legal channels.[27]

Those whom he is defending against are the environmental extremists who accuse this group of having "sold out to the despoilers."

These extremists must be countered, "for in seeking to dismantle modern civilization it has the potential to destroy the very foundations on which a new and ecologically sane order must be built."[28] Lewis goes on to state that he is not personally attacking the deluded eco-radicals and their self-defeating political strategies:

> It is rather their ill-conceived ideas with which I am concerned. If at times my aspersions are caustic, it is because I have had to battle against these seductive ideas myself. Until a few years ago, I too endorsed all of the main platforms of the green radicals.[29]

This wantonness from which Lewis narrowly escapes would hold that

> . . . human society, as it is now constituted, is utterly unsustainable and must be reconstructed according to an entirely different socioeconomic logic. . . . Eco-radicals, therefore denounce anyone seeking merely to reform, and thus perpetuate, a society that they regard as intrinsically destructive if not actually evil.[30]

Eco-radicalism is based on four essential postulates:

> . . . that "primal" (or "primitive") peoples exemplify how we can live in harmony with nature (and with each other); that thorough going decentralization, leading to local autarky, is necessary for ecological and social health; that technological advance, if not scientific progress itself, is inherently harmful and dehumanizing; and that the capitalist market system is inescapably destructive and wasteful. These views, in turn, derive support from an underlying belief that economic growth is by definition unsustainable, based on a denial of the resource limitations of a finite globe.[31]

It is these four postulates which Lewis argues are misguided and deluded, based as they are on

> . . . erroneous ideas fabricated from questionable scholarship. Radical environmentalism's ecology is

> outdated and distorted, its anthropology stems from naive enthusiasms of the late 1960s and early 1970s, and its geography reflects ideas that were discredited sixty years ago. Moreover, most eco-radicals show an unfortunate ignorance of history and a willful dismissal of economics.[32]

Along with these faults, Lewis rejects Herman Daly's contention that ". . . our obsession with economic growth, and more generally, the possession of material goods, has only hidden from our view an underlying spiritual impoverishment,"[33] and is therefore disdainful of any call for a "return to the Earth" that would link human community more closely with natural community.

As opposed to these "radical" approaches, Lewis sets forth what he considers to be a viable approach based on sustainable development, and which endorses capitalism and economic growth:

> Robert Repetto and others at the World Resources Institute argue powerfully that environmental health depends on sustainable development — on ecologically sound economic programs that aim to increase human "wealth and well-being."[34]

Lewis argues that other advocates of sustainable development, such as R. Kerry Turner, stress "the need to view environmental protection and continued economic growth . . . as mutually compatible and not necessarily conflicting objectives."[35] In contrast to this moderate environmentalism which he supports, Lewis states that eco-radicalism not only

> . . . represents a massive threat to human society, to civilization and material progress. . . But the core argument of this work is that green extremism should be more deeply challenged as a threat to nature itself.[36]

By examining the radical attachment to nature and aligning himself with capital reform while encouraging its attendant corporations and regulatory agencies, Lewis clearly places himself in the emergent momentum of development:

> This work accordingly advocates what I call a Promethean environmentalism, one that embraces the wildly creative, if at times wildly destructive, course of human ascent. . . .We should, and we will, continue to burn Prometheus's flame — but we must learn to do so as responsible adults rather than as pyromaniacal adolescents.[37]

Much, I believe, can be made of this statement. The most obvious is that Lewis is affirming in an ascendant frame what I have made problematic in terms of loss. For him, the major aspects of the modern project are not to be called into question and are not in "crisis." He also rejects "subversive postmodernisms" which "deflect" the modern project "wildly" in "radical directions."[38] Environmental problems are therefore not social and cultural problems which are part of a wider crisis of modernity, but are instead technical problems which do not call into question the basic tenets of modernity.

This contrast in views can be best expressed in terms of Lewis' use of "Promethean" to describe his form of environmentalism. If I may be allowed such a statement, Lewis makes Prometheus and Faust brothers who become the "responsible adults" of modernity. This makes my argument concerning the loss of connection between human community and natural community a kind of defense of the "pyromaniacal adolescent," abused in its upbringing within modernity, trapped within a cosmos of past demons, and unable to adjust to the "real" world.

Within this highly divisive and combative milieu which Lewis sets forth, it is my intention to use the sharp contrast that is drawn between moderates and radicals to reveal what I consider to be the representation of a historical moment in the loss of human identification with natural community. The "erroneousness" of the eco-radical arguments portrayed by Lewis is what Williams refers to as a residual form of "a previous social formation"[39] which loses expression because conditions have changed, and the corresponding correctness of Lewis' own views of moderate environmentalism (as allied with emergent globalisation of market economy) are gaining wider expression. So, what Lewis presents as a divergence in views is also a historical moment in which there is a temporary dialogue between what is entering and what is leaving the realm of discussibility.

To illustrate this divergence of human identity and nature (as expressed in the moderates and the radicals), I will list the terms that Lewis uses to describe the two groups. The radicals are of course, "deluded," as well as "seduced" and "shamanistic," have "naive enthusiasms" from which "they denounce" capitalism. They are "mired in idealism," "misanthropic," "disaffected;" they "look longingly" at nature, are "pyromaniacal adolescents" engaged in "anarchic utopianism." They are "credulous" and "blissful," believe in "animism" and "complete pacifism," and are "beguiled by ecoromantic fantasies" — to name but a few of the terms used by Lewis.

The vision of nature described by radicals is an "interconnected totality, the whole of which is greater than the sum of its parts"[40] into which human society must "reintegrate itself." And "we are all fundamentally the same, knots in the vast glorious interconnected web called nature."[41] Their vision of nature is

> one that glorifies the harmonious functioning of undisturbed ecosystems and that considers cooperation among individuals and among species far more common than competition.[42]

Individual elements of nature may undergo change, but "the whole — the marvelous network of interdependencies — is believed to have persisted in perfect harmony, unaltered in its essence."[43] It becomes possible to connect the dislocation experienced by eco-radicals (disaffected, deluded, beguiled) with the corresponding representation of the harmony of nature. The recuperation of this lost harmony into society is central to radical thought, and is the basis upon which Lewis criticizes it as misguided.

By contrast, moderate environmentalists are represented by a very different set of descriptive terms. They are "rational," "logical," "realistic," "pragmatic," "responsible adults," who use a "new economic calculus" to assess human society's use of nature. These humans would operate in a "social, economic and technical milieu almost wholly removed from the intricate web of nature," and would enjoy nature "at somewhat distant remove."

The moderate view of nature is one which weighs "ecological costs" against "economic benefits" so as to insure "environmental stability." Nature is a "scarce resource" and a "public good." And, "if we embrace capitalism, we must accept that it involves . . . creative destruction" of nature. Conservation of nature is seen as making "environmental strides" and instituting "pollution abatement."

One of the most significant contrasts that separates the radical and moderate perspective, is a sense of loss which is present in eco-radicalism and absent from reform minded strategies. Loss seems to be a strange basis for a critique, but it recurs again and again as a central hue which defines the different perspectives of the environmental crisis. To say that the only thing that separates eco-radicals and moderates is a sense of loss, is to risk a kind of reductiveness which would attribute the complexity of a perspective to a single attribute. Nonetheless, I think it is the central battleground over which this divisiveness struggles. This is expressed quite clearly in the contrasting qualities with which Lewis describes moderate and radical conceptions of human identity and nature.

Lewis bases his critique of the "erroneous" aspects of eco-radical conceptions of nature, economics of scale, and technology — a critique which represents the body of the book — by arguing that there is nothing that is fundamentally different about modern culture. This is in sharp contrast to what I have presented as social transformations leading to a crisis of modernity.

What is interesting about the contrasting views in this debate is what is made problematic. For Lewis, it is the assumptions about human attachment to nature which are misguided and a threat to society. For example, in a section entitled "The Threat of Radical Environmentalism," Lewis warns with an almost palpable paranoia:

> Eco-radicalism is admittedly a marginal social movement, its adherents forming an exiguous ideological minority. One might be tempted to conclude that it poses no threat to our economy, our society, and our environment. [Note: "our."] This may ultimately prove true, but it cannot be assumed. Radical environmentalism enjoys substantial, and growing, intellectual clout. If its concerns merge with those of the broader academic left, a trend visible in the rise of both eco-marxism and of a self-proclaimed subversive postmodernism, we may well see the intellectual hardening of uncompromisingly radical doctrines of social and ecological salvation.[44]

Contrast this concern with the undermining of the capitalist consensus to Livingston's description of "threat":

> Entirely out of control, the human technomachine guzzles and lurches and vomits and rips its random crazy course over the face of the once-blue planet, as though some filthy barbaric fist were drunkenly swiping with a gigantic paint roller across an ancient tapestry.[45]

Or, with Dave Foreman's description of human-nature relations and where he locates himself within those relations:

> I am not a machine, a mindless automaton, a cog in the industrial world, some New Age android. When a chainsaw slices into a heartwood of a two-thousand-year-old Coast Redwood, it's slicing into my guts. When a bulldozer rips through the Amazon rain forest, its ripping into my side. When a Japanese whaler fires an exploding harpoon into a great whale, my heart is blown to smithereens.[46]

The divergence in allegiance and the constellation of meaning that is connected with that allegiance, is quite clearly laid out in what I have described as emergent and residual forms within modernity. Lewis presents this divergence in this way:

> Despite Arcadia's sorrows, the dream associated with its name has never died. Especially since Rousseau, disaffected intellectuals have looked longingly to Arcadia as a symbol of the countryside left behind. Unfortunately, not only is Arcadia impossible to reclaim, but it never really existed as imagined in the first place. The Arcadian myth is based on a sanitized picture of nature, one from which labour and suffering have been conveniently removed. If we are to establish a realistic environmental movement we must begin to think in terms of a different Classical archetype.[47]

There is no doubt that there is much that is misguided in the Arcadian view of nature, in the same way that there is much that is misguided in the resourcist view. Current ideas of a pristine nature reflect the displacement of purity elsewhere, which allows for and explains the destruction happening here. Pristine nature and heaven have a lot in common in that regard. Once again, what is obvious here is that — without claiming that nature is definitively one thing or another — the kind of humans we think we are, and the way we understand nature, are closely connected.

Although Lewis claims to support wilderness restoration of the sort advocated by Foreman in the name of this new Promethean archetype, nothing in the rest of his argument — except to say that most times its not even profitable to cut down old-growth forests — promotes a view that would allow for nature preservation or restoration. In fact, for Lewis, this new archetype is most important in its capacities to affirm capitalism, by showing that "labour and suffering" exist everywhere and in every age, and that the modern period is no different than any other.

Despite the denial that his work is based on technological optimism, much of what Lewis has to offer with regard to preserving nature is the general technological and economic wager on the future:

> The Promethean environmentalism advocated here . . . values progress as much for the benefits that it may confer to the human-ravished landscape as for promises it gives to the human community. Although it may seem preposterous, I agree simultaneously with [Julian] Simon on the desirability of technological advance and with Dave Foreman (the archradical founder of Earth First!) on the necessity of wilderness restoration (see Foreman 1991:187). To understand such a seemingly paradoxical position, it is necessary to examine one of this work's fundamental theses: the belief that only by disengaging our economy from the natural world can we allow adequate space for nature itself.[48]

Lewis goes on to state that:

> The prospect of humankind someday coexisting easily with the Earth's other inhabitants — a vision entertained by Arcadian and Promethean environmentalist alike — can best be achieved through gradual steps that remain on the track of technological progress. . . . I believe that only a capitalist economy can generate the resources necessary for the development of a

> technologically sophisticated, ecologically sustainable global economy.[49]

To fulfill this wager:

> More funds must be channeled into education, basic science, and long-term research and development if we are to find an environmentally sustainable mode of existence. While it is essential to guide technology into ecologically benign pathways, it is equally imperative that we consistently support the bases of technological progress itself.[50]

"Support the bases of technological progress itself," is the process within modernity which Lewis sees as the route to environmental salvation. I have posited it as the constellation of processes and relations which have led to the crisis of modernity, and the impoverishment of social relations.

Rather than the archaeology of economic value form as being at the root of the undermining of the connection between human community and natural community as I have argued, Lewis has a very different view of its role:

> It is time to implement a new economic calculus that fully encompasses environmental variables, one such as researchers at the World Resources Institute are presently refining.[51]

It is this kind of understanding about the role of economics which Daly and Cobb[52] single out when they apply Whitehead's conception of "misplaced concreteness" to economic theorizing.

The instrumentalism related to this calculus is not problematic for Lewis. In fact, he accepts as given that this is the path to conservation, rather than any identification with nature because ". . . the public remains, as before, wedded to consumer culture and creature comforts."[53] This leads Lewis to proclaim a call to arms that would conserve nature:

> To galvanize a young electorate whose primary political act to date has been to throw in its lot with capitalism, environmentalism must reconstitute itself with an entirely new philosophical foundation.[54]

This new philosophical foundation is one that embraces "the wildly, creative, if at times wildly destructive, course of human ascent."[55] Central to this new philosophical foundation is a rejection of the eco-radical perspective:

> . . . cynicism is the hallmark — and main defect — of our current age. . . . Although most eco-radicals are anything but cynical when they imagine a "green future," they do take a cynical turn when contemplating the present political order. The dual cynical-ideological mode represents nothing less than the death of liberalism and of reform.[56]

Once again, the threat to the modern ascent does not come from the destruction of nature, but from the "ideologues" who analyse it. As opposed to the processes of capital being at the root of the crisis of modernity, as is argued by the social theorists whom I have cited, Lewis regards these processes as central to the conservation of resources.

Accentuating the Gulf

> In a Promethean environmental future, humans would accentuate the gulf that sets us apart from the rest of the natural world — precisely in order to preserve and enjoy nature at somewhat distant remove.
>
> *Martin Lewis*[57]

> . . . money and the enlargement of its role places us at an increasingly greater mental distance from objects. This often occurs in such a way that we lose sight of their qualitative nature so that the inner contact with their whole distinctive existence is disrupted. This is true not only of cultural objects; our whole life has become affected by its remoteness from nature, a situation that is reinforced by the money economy and the urban life that is dependent on it.
>
> *Georg Simmel*[58]

I have spent much of this work trying to chart the modern transformation of human society related to changes in the money form, in order to show the profound differences between the way modern society understands human-nature relations, and Livingston's description of unperturbed natural community. By contrast, Lewis devotes the body of his book to a dismissal of any significant alternative to capitalism. In this, Lewis universalizes present realities by showing that "labour and suffering" exist everywhere. Lewis engages in this presentation of history, intent on showing that the longing look of eco-radicals to a harmonious Arcady is misguided and never existed anyway. Whether or not there is any credence to the Arcadian view of nature, the intention here it seems is to create a frame of reference for the discussion of environmental problems that is firmly based on a modern capitalist world view. Whereas I have struggled to create a frame of reference to which "the market itself is referable," Lewis' main goal is to erase that referentiality.

In a section entitled "Eden Revisited: Scholarly Miscues," it is this kind of universalizing that Lewis accuses eco-radicals of engaging in:

> . . . radical environmentalism's vision of primal harmony is so exaggerated as to verge on intellectual fraud. Its fundamental error is overgeneralization. Eco-radicals seldom distinguish among tribal groups, picturing them instead as the undifferentiated "other." The result is that, despite acknowledgment of superficial diversity, primal human-environmental relations are seen as being essentially the same everywhere. "For the *primal mind* there is no break between humans and the rest of nature" [Devall and Sessions 1985:97; emphasis added] [Reference by Lewis].[59]

Lewis presents the Eden story as an example of a false attachment to unity, which is also engaged in by eco-radicals. My contrasting analysis represents the harmony of the Eden story as a record of loss that is deeply connected to the expulsion into new Neolithic ways of relating to nature. The expression of primal unity within the context of loss is a reflection of that grief and loss. So, rather than it being a kind of false nostalgia, as Lewis argues, these harmonic representations — misguided or not — are instead a confirmation of what I have argued as the residual cultural record of loss of identification with nature that has occurred throughout human history. There is no doubt that these representations of unity have far more to do with human aspiration than with a viable representation of natural community. But the significance of them also has much to do with the realities of capital and markets than the criticism of "scholarly miscues" implies.

Throughout his critique of "primal purity," Lewis attempts to show that there indeed never was any purity, and that the reality of those supposed alternatives put forth now by eco-radicals, are not much different than the modern one: ". . . it is clear that the world of the Yanomamo is anything but peaceful."[60] Lewis goes on to state that "if I have presented a rather tiresome parade of human iniquities, it is only to show only that the human destruction of nature and the exploitation of fellow humans are facts of *long standing.*"[61] This conception of the relation between humans and nature leads him to conclude:

> If we are to construct an environmental movement powerful enough to enact needed reforms we must first relinquish our romantic fantasies. A meaningful environmentalism cannot be based on nostalgia.[62]

The disembedding processes of capital are not "long-standing," but are associated with particular processes of modernity. Whereas I have represented loss or "nostalgia" as an avenue into a critique of assumptions which underpin modern conservation discourse, Lewis affirms that there is nothing different about modernity, and that a recognition of loss of human identification with nature is a misguided Romantic fantasy.

In the same way that Brown[63] conceives of the presence of the devil in early modern history as a representation of the loss related to modern transformations into "spurious rationality" which appears momentarily in the work of such men as Luther, so the "maladjustment" of eco-radicals and their sense of loss, is a temporary appearance that acknowledges a change in social and economic relations (which Lewis' work would render invisible). The call for the abandonment of nostalgia is a confirmation of an ascendant trajectory related to an intensification of the globalisation of the world economy.

Similar to the primal purity argument that all cultures are the same, Lewis' chapter entitled "A Question of Scale" argues that there is no real separation between country and city: "Each [has] created the other, so their mutual transformations in fact [express] a single system and a single history."[64] This "single history" leads

Lewis to make such statements as, "preindustrial peoples have proved themselves capable of extraordinarily destructive acts, notably by deforesting entire landscapes and exterminating major faunal species"[65] and, "as I have endeavoured to show, the West holds no monopoly on environmental destructiveness. In fact, in many respects the East is guilty of the modern world's most extreme violations."[66] Lewis sums up this single history with this admonition:

> . . . we must realize that destructiveness and exploitation have been features of all societies. To reduce the evil in human history to the functioning of a specific economic form (capitalism), or to limit it to a single subcontinent (Europe), is a foolish gambit indeed.[67]

Although I would not try to associate capitalism in any deterministic way with a specific culture or region, the main argument of this work is that transformations in the money form — of which the development of capital in the West in the modern period is the most dramatic expression — are at the heart of the current "destructiveness and exploitation" of the natural world.

After the discussion which concludes that all societies are like modern society, Lewis then moves on to a vision of his environmental future which he titles "The Capitalist Imperative." He begins by challenging the eco-radical conception of capitalism:

> [T]he most immediate denunciations of capitalism focus on its profit-seeking orientation. Capitalist firms are seen as avaricious in their pursuit of gain, ruthlessly exploiting workers, consumers, and the land so owners can accumulate ever growing piles of wealth. . . the logic of capital alone is thought to demonstrate the system's exploitative nature. . . .[68]

In response to this criticism of capitalism by eco-Marxists and others, Lewis replies that, in light of the collapse of communism and the "spectacular success" of capitalist economies, "the thesis that capitalism is destined to fail from its own internal contradictions is a bit threadbare these days."[69] Lewis also claims that the signs of doom for modern society have been "overestimated." I have argued throughout this book that the logic of money and capital do indeed "demonstrate the system's exploitive nature," as well as frame the way we understand those exploitation patterns. Through the massive inversion of reality that drives capital, we increasingly grant social standing to dead things (commodities) and deny it to living things (humans and nature).

After dismissing the eco-radical critiques of capitalism, as well as denying any viable alternative to it, Lewis goes on to lay out what he sees as a solution based on a "guided capitalism" which promotes a

> . . . corporate and market system in which the state mandates public goods, in which taxes are set to level social disparities and to penalize environmental damage, and in which fiscal policies are manipulated to encourage long-term investment in both human and industrial capital.[70]

Why is this statement so eminently sensible to moderates and so appalling, even unthinkable, to those of the eco-radical perspective? On what basis, and for what reason, would someone who conceives of human identity in terms resources and capital, begin to find enough significance in the complex sociality of wild beings to care about conserving them? It is clear that by using the phrase "human and industrial capital" — two adjectives that accompany the noun — Lewis has made human identity the operand of industrial production. Community has gone from being a verb to a noun, and now an adjective; one instance among many in the pervasive categories of capital which anchor understanding. Any viable discussion of the relationship between human community and natural community becomes inexpressible in this kind of world.

If — as I contended earlier in the chapter — discussions about the environment are essentially struggles over frames of reference in which very different issues are made problematic, these kinds of questions do not appear to be of any concern to Lewis. Instead, he develops a view of society in which

> . . . an external referee — the state — is necessary just as it is in sports. Corrupt or incompetent state offi-

> cials can still fail at their duty, but this calls for greater vigilance, not repudiation of the system.[71]

Within this context,

> The role of the government is thus not only to intercede between firms to ensure healthy competition, but also to encourage economic growth and to provide public goods — including environmental protection.[72]

In the same way that human community is remade as only one point on a continuum of resources available to capital, natural community appears only vaguely and tangentially under the term "environmental protection."

The closest Lewis can come to nodding in the direction of other sentient beings besides humans, is to state that

> . . . we need both an overriding ethical system that accords intrinsic worth to nonhumans — without insisting on pan-species egalitarianism — and an efficient means of allocating scarce resources.[73]

Lewis does not pursue the philosophical ramifications of acknowledging the intrinsic worth of scarce resources while they are being allocated efficiently. It is this kind of statement by Lewis which the work of social theorists render as the impoverished categories of capital and markets in which humans and nature are fictitiously commodified.

The contrasting perspectives of moderate and radical environmentalism described by Evernden in *The Natural Alien* also reflect this divergence between the person who accepts science and prediction as a basis for the "conservation" of resources, and the one who in subjective terms can "attest to his own experience of a meaningful, valuable, colorful world" in which humans are inextricably included.[74]

In contrast to Lewis' attempt to argue that what is going on currently with regard to the destruction of the natural world by industrial society is part of a "single history" which is not that different from other historical periods or other cultures, Evernden highlights the profound changes in the relationship between human society and nature which have made modern humans into exotic creatures:

> The consequences of technology are subtle but extensive, and one such consequence is that man cannot evolve *with* an ecosystem anywhere. With every technological change he instantly mutates into a new — and for the ecosystem an exotic — kind of creature. . . [and] just as the environment does not know how to cope with the new creature, neither does the exotic know what to do. In other words, the exotic is a problem because it does not know how to comply.[75]

What is so profoundly different and so predatory about this modern exotic is that the world is divided into just two things: "humans and natural resources." As opposed to the compliant behaviour of natural beings which promotes the success of the community as a whole, the modern human exotic bases its success, to quote Evernden again, "on the ability *not* to care, to remain dispassionate, unattached, and objective."[76] Or in Lewis' terms, to "decouple" human society from nature so that it can be enjoyed at "somewhat distant remove."

Rather than this dispassionate remove leading to the conservation of what is left of nature — as Lewis holds — Sachs sees it in terms of the completion of the domination of nature by humans:

> . . . the responsiveness of nature has been strained to the uttermost under the pressures of modern man. Looking at nature in terms of self-regulating systems, therefore, implies either the intention to gauge nature's overload capacity or the aim of adjusting her feedback mechanisms through human intervention. Both strategies amount to completing Bacon's vision of dominating nature, albeit with the added pretension of manipulating her revenge. In this way, ecosystem technology turns finally against ecology as worldview. A movement which bade farewell to modernity ends up welcoming her, in new guise, through the backdoor.[77]

It is this monolith which Lewis would create by banishing the insights of radical environmentalism. In this, his work attempts to complete the work of the processes of development. In the same way that it was impossible to be an atheist prior to the seventeenth century in Europe, it may also be impossible to be an "eco-radical" in coming generations. In Lewis' decoupled world, it would be unimaginable.

NOTES

1. Karel Kosik. 1976. *Dialectics of the Concrete*. Boston: D. Reidel Publishing, p. 41.
2. Martin Lewis. 1992. *Green Delusions: An Environmentalist Critique of Radical Environmentalism*. Durham: Duke University Press.
3. Robert Heilbroner. 1985. *The Nature and Logic of Capitalism*. New York: Norton, p. 20.
4. Maurice Strong. 1992. *Toronto Star* (October 30, 1992:A9)
5. Ruth Grier. 1992. *Toronto Star* (September 23, 1992:A3)
6. International Union for the Conservation of Nature, United Nations Environment Programme, World Wide Fund for Nature. 1980. *World Conservation Strategy*. Gland Switzerland, Section 8.
7. World Commission on Environment and Development. 1987. *Our Common Future*. New York: Oxford University Press, p. 37.
8. J. MacNeill, P. Winsemius, and T. Yakushiji. 1991. *Beyond Interdependence*. New York: Oxford University Press, p. 4.
9. Robert Costanza. 1991. *Ecological Economics: The Science and Management of Sustainability*. New York: Columbia University Press. p.9.
10. World Conservation Union, United Nations Environment Programme, World Wide Fund for Nature. 1991. *Caring For the Earth*. Gland, Switzerland, p. 8.
11. Dave Foreman. 1991. *Confessions of an Eco-Warrior*. New York: Harmony.
12. Stephen Marglin. 1990. *Dominating Knowledge*. Oxford: Clarendon, p. 2.
13. Jean Baudrillard. 1975. *The Mirror of Production*. St. Louis: Telos Press, p. 143.
14. Marglin (1990:16).
15. Baudrillard (1975:143).
16. Wolfgang Sachs. 1992. *The Development Dictionary*. London: Zed Books, pp. 35-36.
17. Scientific American. 1990. *Managing Planet Earth*. New York: Freeman, p. 110. James MacNeill. 1990. "Strategies for Sustainable Economic Development." in Scientific American's *Managing Planet Earth*.
18. H. Jeffrey Leonard. 1985. "Introduction," to *Divesting Nature's Capital* edited by H. Jeffrey Leonard. New York: Holmes and Meier, p. 3.
19. Colin Clark. 1990. *Mathematical Bioeconomics*. New York: Wiley, p. 68.
20. Robert Repetto. 1986. *World Enough and Time: Successful Strategies for Resource Management*. New Haven: Yale University Press, p. 128.
21. Henry Thoreau. 1962. *Walden*. New York: Houghton Mifflin.
22. Martin Lewis. 1992b. "Green Threat to Nature." Harper's Magazine. November, p. 26.
23. Lewis (1992b:26).

24. Lewis (1992b:26).
25. Lewis (1992:16).
26. Walter Benjamin. 1972. *Illuminations*. London: Collins, p. 257.
27. Lewis (1992:2).
28. Lewis (1992:2).
29. Lewis (1992:2).
30. Lewis (1992:2).
31. Lewis (1992:3).
32. Lewis (1992:11).
33. Lewis (1992:4).
34. Lewis (1992:4).
35. Lewis (1992:4).
36. Lewis (1992:6).
37. Lewis (1992:15).
38. Lewis (1992:161).
39. Williams (1980:40).
40. Lewis (1992:28).
41. Lewis (1992:29).
42. Lewis (1992:43).
43. Lewis (1992:45).
44. Lewis (1992:247).
45. John Livingston. 1981. *The Fallacy of Wildlife Conservation*. Toronto: McClelland and Stewart, p. 20.
46. Dave Foreman (1991:5).
47. Lewis (1992:15).
48. Lewis (1992:16).
49. Lewis (1992:18-19).
50. Lewis (1992:19-20).
51. Lewis (1992:20).
52. Herman Daley and John Cobb. 1990. *For the Common Good*. Boston: Beacon Press.
53. Lewis (1992:11).
54. Lewis (1992:14-15).
55. Lewis (1992:15).
56. Lewis (1992:248).
57. Lewis (1992:18).
58. Georg Simmel. 1990. *The Philosophy of Money*. New York: Routledge, p. 478.
59. Lewis (1992:55).
60. Lewis (1992:71).
61. Lewis (1992:81).
62. Lewis (1992:81).

63. Norman O. Brown. 1985. *Life Against Death: The Psychoanalytical Meaning of History*. Middletown: Wesleyan University Press, p. 213.
64. Lewis (1992:101).
65. Lewis (1992:149).
66. Lewis (1992:244).
67. Lewis (1992:246).
68. Lewis (1992:152).
69. Lewis (1992:167).
70. Lewis (1992:20).
71. Lewis (1992:177).
72. Lewis (1992:178).
73. Lewis (1992:180).
74. Neil Evernden. 1985. *The Natural Alien*. Toronto: University of Toronto Press, p. 33.
75. Evernden (1985:109).
76. Evernden (1985:111).
77. Lewis (1992:32).

Conclusion
Horizons of Significance

In *The Ethics of Authenticity,*[1] Charles Taylor discusses what he considers to be a loss of meaning in modern culture. Taylor describes a human-centred perspective which he links with liberal neutrality and moral subjectivism. This has led to the erosion of human meaning, or horizons of significance that threatens to trivialize life in modern society. Taylor associates anthropocentrism, or human-centredness, not so much in biological terms as environmentalist do when they talk about the way humans regard the rest of the world as a store-house of resources for their use, but in terms of an anti-metaphysical, anti-spiritual humanism which has led to an increasingly "flattened world, where the horizons of meaning become fainter." "We face," Taylor states, "a continuing struggle to realize higher and fuller modes of authenticity against the resistance of the flatter and shallower forms."[2]

Taylor's example of this shallower form is the relationship between human society and the natural world as it is represented in terms of an environmental policy which recommends restraint in the use of resources. In contrast to the merely pragmatic and instrumental appeal to save the forests and the seas for the sake of our descendants' happiness, Taylor insists that "we need to see ourselves as part of a larger order that can make claims on us." This connection to a larger order is to be found in works that are in opposition to what I have characterized as development:

> It was perhaps not an accident that in the Romantic period the self-feeling and the feeling of belonging in nature were linked. Perhaps the loss of a sense of belonging through a publicly defined order needs to be compensated by a stronger, more inner sense of linkage. Perhaps this is what a great deal of modern poetry has been trying to articulate; and perhaps we need few things more today than such articulation.[3]

I would add to Taylor's statement by contending that the strength and intensity of the Romantic linkage with nature is connected with its loss of expression within the changing "publicly defined order." Or, to repeat Paul Shepard's statement, ". . . the real bitterness of modern social relationships has its roots in the vacuum where a beautiful and awesome otherness should have been encountered."[4] This vacuum is the financial vortex of modern economics.

I would argue that this contrast between the articulation of a "loss of a sense of belonging" and the shallow and custodial maintenance of resources, is also at the heart of Lewis' description of the difference between radical and moderate environmentalists. Although Lewis claims to promote a guided capitalism in the interest of conserving nature, there is nothing in his perspective that places humans "as part of a larger order that can make a claim on us." In fact, his approach is an instrumental one in which "human and industrial capital" make use of a "new economic calculus" to regulate the relationship between human society and nature. By contrast, eco-radicalism is based first and foremost on claims that the rest of nature can make on humans.

In order to focus on Lewis' new philosophical foundation of environmentalism in terms of what I have argued with regard to the creation of cultures of substitution related to transformation in money forms, I will conclude my discussion by outlining the conception of human identity which Lewis puts forth as central to this new philosophy. In a section entitled "The Reinvention of Bourgeois Values," Lewis states:

> Perhaps the creation of an environmentally benign economic order calls for a return to, or an invention of, a truly capitalistic ethos. Capital itself must be regarded as virtually sacred — it represents nothing less than the savings necessary to construct a more prosperous and less environmentally destructive future economy. Capital is deferred gratification writ large. Just as a child must learn to delay the satisfaction of his or her wants in order to become a responsible adult, so too the public must learn to put away for the future.[5]

Scarcely a word of this quotation — which for Lewis represents a positive initiative — does not at the same time reverberate back through the arguments I have made with regard to the contributing factors leading to the crisis of modernity, and the destruction of the natural world. A "truly capitalist ethos" that is "environmentally benign" runs counter to everything I have argued with regard to modern economics. The "expelling into transcendence" of sacred capital has been central to the critiques by Marx, Polanyi, Simmel, and Brown, and which chart transformations in the money form and the related alienation of relationships. With reference to Goux and Baudrillard, this "sacredness" of meaning separate from the embedded relationships of the things themselves, becomes the generalized condition of late modernity leading to a "crisis of modernity" (as human society operates increasingly within a self-referential code based on substitutive processes).

The point of this work is to confirm that indeed "capital is deferred gratification writ large," and to relate that social construction to both the collapse of viable human identities and viable arguments for the conservation of nature. To advocate as Lewis does, that modern humans stop behaving as children and "to put away for the future," is to proclaim the triumph of death in life. "Life" is displaced and canonized — as the Calvinist burghers had done — into an immortalized "expelled" future existence, and nature is expelled from the human context to a wilderness area where it can be appreciated at somewhat distant remove. To defer to the future is not to be a responsible adult, as Lewis argues, but to remain forever unmade.

Livingston presents a similar critique of shallow anthropocentrism to Taylor's, but his perspective relates to a ecological "claim" on humans, rather than a metaphysical one:

> We became no longer the masters and users of abstraction, but its dependents. We have moved into a state of utter domesticated servitude to instrumental reasoning, and in so doing have lost (or abrogated) our participatory relatedness to the wider life community. No longer able to perceive the rest of nature in any way other than the abstract and utilitarian, we have become ecological loose cannons, dangerous misfits in what had been theretofore a fully integrated and mutually compliant and supportive biosphere.[6]

Livingston compares this sacred saving for the future, with the normal patterns of comings and goings in natural communities:

> A difficulty arises [for humans] in the fact that by definition the future never comes. Our self-identity is

> never found. If we identify solely with the future, then surely we never grow up. We are forever becoming, forever questing for mature adulthood.[7]

For both modern humans and nature, management replaces maturation. In this context, I would argue that conceptions of sustainability have less to do with conservation, and more to do with striving for immortality. Modern humans manage the Earth, and they cannot stand the idea of not doing it forever. So they develop the idea of sustainability to add an immortal aspect to exploitation.

Lewis dismisses small scale societies such as the Melanesians as covetous, and then defines an ideal of human identity that would be central to this new environmental archetype:

> The best examples of voluntary simplicity may be found in capitalist societies, from groups such as the Calvinist burghers of the early modern period to the Japanese "salarymen" of the present.[8]

This ideal set forth by Lewis resonates back to Taussig's definition of fetishism:

> Fetishism denotes the attribution of life, autonomy, power, and even dominance [Lewis' sacred capital] to otherwise inanimate objects and presupposes the draining of these qualities from the human actors who bestow the attribution.[9]

Brown, Marx, and Simmel have made the significance of the Calvinist burgher central to the development of modern political economy and what I have argued are processes leading to a set of relationships defined as cultures of substitution. None of this kind of analysis is problematic for Lewis as he lays out the human ascent through the modern period that arrives at the lofty archetype of the "salaryman."

It is not my intention here to reject everything that Lewis has argued. In fact, next to total ecological collapse, his presentation of the future is the most likely path modern society will follow. His work then, is a confirmation that humans indeed no longer need nature as part of their identity. The environmental crisis is not solved in works such as that of Lewis. Instead, human identity is remade in accordance with the new realities, and life goes on. At least for humans it goes on, even if they have to purchase fresh air from a street vendor to do so. The same cannot be said for the rest of nature, which is imperiled on every side.

The central issue here — and the one which this book has tried to confront by framing its question in terms of the connection between the disappearance of the natural world in a real sense, and the concurrent disappearance of the human impetus to conserve nature — is that arguments related to conserving nature are inextricably linked to representations of human identity. What is required to be "human" will be expressed in corresponding conceptions of the nature which is recognized as being significant. The question that has to asked is, given the qualities of the "salaryman" that are required of a human identity in Lewis' future, can that identity find a reason to preserve the qualities of nature outlined by naturalists such as Livingston? In this, any argument related to the conservation of nature must similarly link humans and nature. In other words, conservation is about social community. For Lewis, by contrast, it is a matter of "human capital" looking for "intrinsic worth" in "scarce resources." What is clear about Lewis' conceptualization of environmental issues is that both humans and nature are being "managed" within the categories of efficiency and profit, even if there are conservation measures which would allow for future use. Capital will tolerate no other community.

What is significant about *Green Delusions* is that it attempts to articulate in a philosophical sense, what is usually implicit in conservation tracts which focus on the integration of conservation and development under the name of sustainability. By examining the "erroneousness" of eco-radical philosophy, Lewis allows that approach to define the terms of discussion. That makes it an unusual book and an exceedingly useful one for an analysis of the struggle about frames of reference, because it leads Lewis to set forth the general assumptions about human identity and nature which inform moderate environmentalism. What is clear about these assumptions is that they are in no way different from traditional capitalist conceptions, and that any movement towards the conservation of nature through this moderate approach, is achieved in a pragmatic reform of the modern tools at hand.

In his dismissal of the eco-radical statements about the failure of capitalism to conserve nature, Lewis has been very selective in what he has found problematic about modernity. Like the Calvinist Burgher's ascendant view of realities which overrode the grief and loss of Tragedy, or the industrialist's overriding of

Romanticism, Lewis' vision of moderate environmentalism is located within the emergent transformation of global economics and management, and renders invisible the insights of radical environmentalism. The array of perspectives which I have chronicled, and which identify a contemporary crisis of modernity related to the evolution of the money form — of which the environmental crisis is one expression — are not problematic in Lewis' work except as a threat to a new world order. Instead, he redefines the constituents of the world which would make this threat inexpressible.

From within the closure of Lewis' vision of the future, it is not so much that the "shamanistic" emotions of radical environmentalism would disappear, just as the emotions of Tragedy or Romanticism have not entirely disappeared from Western culture. But I do contend that these emotions are no longer an intricate part of human identity, and that — like Tragedy and Romanticism — eco-radical perspectives will be periodized as the cultural expression of "a way of life losing its life." Where there was once an identification between humans and nature, there is now a new calculus of resource management.

The Aftermath of Collapse

Last summer, I was up at the gas station near my home in Little Harbour, Nova Scotia getting the oil changed in my truck. As Vernon Murphy, the mechanic, stood under the hoist watching the black glistening liquid stream down into a discarded plastic pan from one of the recently-closed local fish plants, he told me about the time he was longline fishing out on the edge of Roseway Bank with Hal Harding.

"I was just a kid, maybe 14 or 15. We had hauled our nets in the night on the way out to the bank, and we got a couple a hundred pounds of herring, enough to bait out five or six tubs of gear. We were in range of the bank about four o'clock and were 'laying to' while Hal lowered the sounding lead to see what depth we were in and I was chopping up the herring for bait. Usually we chop it up as we steam out to the bank, but there was an old southwest slop in the water and it was too rough. A couple of baits fell over the side as I was chopping and a school of squid came up to get'm. I grabbed a bucket or a broom or whatever was handy and scooped as many of the squid aboard as I could."

"'Come on boy,'" said Hal in that high squeaky voice of his. "'If you keep fooling around back there we'll never get this gear out by daylight.'"

"Bye and bye, I managed to get a bucket full of squid and I chopped them up and baited nearly a tub of gear on it. Well look'ee here son, we were hauling along on that herring the next morning and hook after hook came up empty. We were maybe getting a hundred pounds on six lines of gear, mostly trash — small cod and haddock. Then we came to the squid I'd baited on, and the line was smother white coming up. There were steak cod as big as men as far down as you could see. Not a hook missed, and the hooks were only six feet apart in them days."

"Why sonny," said Hal in that voice of his as he looked down into the water, "' if you would of chopped those squid baits the size of your thumb nail, we would of loaded her.'"

"And him cursing me the whole time for taking the time to scoop'm aboard. I remember he sunk the gaf into one of those steakers and rolled him up on the rail — its head the size that wheel rim there — he looked back at me with a glint in his eye and said, 'It's moments like this, I know who I am.'"

This book has attempted to ask the question of modern society, "who are we?" or "what have we become?" by examining other moments in modernity. But what does my presentation of resistance and refusal mean for the coastal fishing communities of Atlantic Canada? Should they now read Keats and Shakespeare and become eco-radicals? I do not think they will do that, but what they do have to do is to find a way to understand themselves and their relationship to the wider marine community in ways that are not entirely compromised by the intellectual ruin of development. This is a social and cultural project, rather than a management initiative.

This approach leads to the conclusion that in order to solve environmental problems, it is necessary to solve history. A few people have tried to solve history and have not had much success. What I mean by "solving history" is that the social, economic, political, and cultural structures which permeate everyday life, are also the structures which are causing environmental problems such as the collapse of the fishery. The autocracy of the present makes rethinking the human condition all but unthinkable. I therefore feel that the idea of solving environmental problems in a limited instrumental sense within the current dominant paradigm offers little hope when compared with the destructive capabilities which remain in place.

Modern humans are at risk of becoming 'one note wonders' thumping out their dreary anthem to capital. This is not a very promising basis for even asking the right questions, let alone coming up with any answers. I am not saying that all is lost or that modernity is one single monolith with no divisions within it. What I am saying is that given the dominant perspective which is now moving us toward globalisation of the world economy, I see no hope. Any initiative towards conserving the natural world is overridden ten-fold by pressures of over-exploitation, and usually only begins in the aftermath of economic and ecological collapse.

Where I do see some hope is in the resistance of groups which still have some form of embedded relations — to use Polanyi's term — with their surroundings. As I stated in the introduction, conservation is not an on/off switch for destructive behaviour imposed at some upper level by an external authority. Conservation has to be implicit in the activity. In other words, humans have to recognize that they are members of not just human communities, but also natural communities. To protect this before it is lost is obviously much easier than recreating it self-consciously in the aftermath of collapse. This is because conservation is not an isolated instrumental problem, but is connected to everything.

For those who live at the heart of modern culture, an important first step towards conserving the natural world is the recognition of the massive inversion which has taken place, and which has granted social standing to commodities and withdrawn it from social communities. Akin to the feminist call to "Take Back the Night," humans have to reclaim what they have lost to the de-animating processes of capital. By recognizing that capital is "only" a set of relations which have, after all, been created by humans, they may then set about changing those relationships.

So what does this mean for Atlantic communities, human and natural? In seeking the dream of modernity, they are now living out its nightmare. Cut off from their surroundings and dependent on that more fickle source of life — the federal government — they are entreated to leave, to retrain, or to professionalize and upgrade their status as fishers. The mind boggles when one considers that the last major initiative of the Federal Department of Fisheries and Oceans before it embarked on the current shutdown, was to professionalize the bona fide fisher category. With everything coming down around their ears, functionaries in the federal bureaucracy were — and still are — preoccupied with developing criteria for upgrading the navigational and fishing skills of those in the fishery, so that they can be assessed in terms of their professional standing in the industry. The unspoken goal of this program is to turn fishers into entrepreneurs and professionals so that the fishery would no longer be a community problem, but instead a business management problem which the market would take care of, the way it takes care of all business problems. In other words, make capital the only community.

In contrasting the local and the global, it is not just a matter of differing scales and carrying capacities. It is also a difference in understanding. Returning to my original thesis statement that not only is the natural world disappearing in a real sense, but the modern human impetus to conserve what is left of nature is also disappearing, it is important to note once again that this disappearance is related to the standardizing of the increasingly global playing field of capital. Affirming a local perspective is to highlight the transformation in relationships which has occurred in conjunction with this standardization. To affirm a particular place is to struggle to find a way to talk about global capital, in the same way that it is necessary to create the idea of diversity in order to talk about monocultures. In this sense, local diversity and global monoculture are inextricably linked, and are representative of modern discourse. Cultures based on local diversity would not recognize themselves in those terms because diversity requires homogeneity against which to define itself. In other words, local diversity becomes discussible when it becomes problematic.

So, when I say that a possible first step to solving the problems of the fishery is a community based one, I am affirming a perspective as much as a place. It is therefore not some authentic basis that is returned to, but a self-conscious creation of a new place which is in opposition to capital. Such is the life of the exotic. There is, of course, no such new place for natural communities that could save them from the loss of their original habitat.

When I began fishing, the largest task — aside from getting over the dread of venturing out into the North Atlantic night alone — was learning about the ocean bottom and the places where the fish might be. I spent a lot of time poring over the navigational charts and plotting courses and times in search of the mother lode. I would then take the chart down to the wharf and seek out Arthur Swansburg, a seventy-year-old fisherman who had spent his life on the waters which I was now venturing into. Spreading the chart out on the hood of the truck, I asked Arthur about the various places that I thought about going to that night. He would begin to describe each place, waving his hand in a vague way over the chart. I was hunched over, following his directions along the contour intervals with my finger. "You head Sou'west keeping her in 40 fathom, until she begins to drop off to 45 fathom. Then you haul West-Sou'west or you will end up in the mud."

I looked up and realized that Arthur and I were having two different conversations. Arthur had turned around and was looking out to sea. He had his eyes closed and was feeling the wind in his face as he was describing the place to me. "If the tide is flooding you'll have to be careful your gear doesn't get swept down over the edge." He had to imagine the place and put himself there. The abstracted perspective of the navigational chart meant nothing to him.

On the voyage out, I made good use of what Arthur had told me. On the voyage back, we have sailed into a harbour neither of us recognize. We will have to imagine a home and put ourselves there.

NOTES

1. Charles Taylor. 1991. *The Ethics of Authenticity*. Cambridge: Harvard University Press.
2. Quoted in review by Richard Rorty entitled "In a Flattened World," *London Review of Books*, Vol. 15, No. 7, April 8, 1993, p.2.
3. Taylor (1992:91).
4. Paul Shepard. *Nature and Madness*. San Francisco: Sierra Club, p. 108.
5. Martin Lewis. 1992. *Green Delusions: An Environmentalist Critique of Radical Environmentalism*. Durham: Duke University Press, p. 187.
6. John Livingston. 1992. "Les implications écologique du caractère impératif de l'immortalité de l'homme," in *Frontiéres*, Vol. 5, No. 2, Automne, p. 9.
7. Livingston (1992:9).
8. Lewis (1992:188).
9. Michael Taussig. 1980. *The Devil and Commodity Fetishism in South America*. Chapel Hill" University of North Carolina press, p. 31.

Bibliography

Abrams, M. H. 1971. *Natural Supernaturalism*. New York: Norton.

Appadurai, A. [ed.] 1988. *The Social Life of Things*. Cambridge: Cambridge University Press.

Barber, Benjamin. 1991. "Enough about rights. What about Obligations?" *Harper's Magazine,* February. p. 49.

Bateson, Gregory. 1972. *Steps to an Ecology of Mind*. New York: Ballantine Books.

Baudrillard. Jean. 1975. *The Mirror of Production*. St. Louis: Telos Press.

Benjamin, Walter. 1973. *Illuminations*. London: Collins.

Berger, Harry. 1988. *Second World and Green World*. Los Angeles: University of California Press.

Berman, Marshall. 1988. *All That Is Solid Melts Into Air*. New York: Penguin.

Berman, Morris. 1981. *The Re-Enchantment of the World*. Ithaca: Cornell University Press.

Birch, Thomas. 1991. "The Incarceration of Wildness: Wilderness Areas as Prisons" in *Environmental Ethics,* Winter.

Bloom, Harold and Lionel Trilling. [eds.] 1973. *Romantic Poetry and Prose*. New York: Oxford University Press.

Bohannan, Paul. 1959. "The Impact of Money on an African Subsistance Economy." *The Journal of Economic History*. Vol. XIX, No. 4, pp. 491-503.

Brenner, Robert. 1985. *The Brenner Debate*. New York: Cambridge University Press.

Bromley, D. and M. Cernea 1989. *The Management of Common Property Resources*. Washington: World Bank.

Brown, Norman O. 1985. *Life Against Death*. Middletown: Wesleyan Univ. Press.

Burke, Peter. [ed.] 1972. *Economy & Society in Early Modern Europe.* New York: Harper & Row.

Canada. 1974. *Law of the Sea Discussion Paper.* Ottawa: Department of External Affairs.

Cayley, David. 1991. *The Age of Ecology.* Toronto: Lorimer.

Clark, Colin. 1990. *Mathematical Bioeconomics.* New York: Wiley.

Costanza, Robert. 1991. *Ecological Economics.* New York: Columbia University Press.

Crump, T. 1981. *The Phenomenon of Money.* Boston: Routledge.

Dalton, George. 1961. "Economic Theory and primitive Society," in *American Anthropologist*, No. 63, pp. 1-25.

Daly, Herman and John Cobb. 1990. *For the Common Good.* Boston: Beacon Press.

Debord, Guy. 1983. *Society of the Spectacle.* Detroit: Black and Red.

Dollimore, Jonathan. 1984. *Radical Tragedy.* Chicago: Univ. of Chicago Press.

Drache, Daniel and Meric S. Gertler. [eds.] 1991. *The New Era of Global Competition.* Montreal & Kingston: McGill-Queens University Press.

Eagleton, Terry. 1986. *William Shakespeare.* Oxford: Basil Blackwell.

Ehenfeld, David. 1978. *The Arrogance of Humanism.* New York: Oxford University Press.

Eliade, Mircea. 1959. *The Sacred and the Profane.* New York: Harcourt Brace Jovanovich.

Evernden, Neil. 1992. *The Social Creation of Nature.* Baltimore: The Johns Hopkins University Press.

Evernden, Neil. 1985. *The Natural Alien.* Toronto: University Of Toronto Press.

Fisheries and Marine Service. 1976. *Policy for Canada's Commercial Fisheries.* Ottawa: Department of the Environment.

Foreman, Dave. 1991. *Confessions of an Eco-Warrior.* New York: Harmony.

Foucault, Michel. 1973. *The Order of Things.* New York: Vintage.

Galbraith, John. K. 1975. *Money: Whence It Came, Where It Went.* Boston: Houghton Mifflin.

Gaull, Marilyn. 1988. *English Romanticism: The Human Context.* New York: Norton.

Glacken, Clarence. 1967. *Traces on the Rhodian Shore: Nature and Culture in Western Thought from Ancient Times to the End of the Eighteenth Century.* Berkeley: University of California Press.

Goldmann, Lucien. 1964. *The Hidden God.* New York: Norton.

Goux, Jean- Joseph. 1990. *Symbolic Economies.* Ithaca: Cornell University Press.

Guggenheim, Thomas. 1989. *Pre-capitalist Monetary Theory.* New York: Pinter.

Hardin, G. and J. Baden. 1977. *Managing the Commons.* San Francisco: Freeman.

Harries-Jones, P., A. Rotstein, and P. Timmerman. 1992. "Nature's Veto: UNCED and the Debate Over the Earth" University of Toronto: Science for Peace.

Heilbroner, Robert. 1985. *The Nature and Logic of Capitalism.* New York: Norton.

Hill, Christopher. 1974. *The World Turned Upside Down.* Harmondsworth: Penguin.

Hummel, Monte. [ed.] 1989. *Endangered Spaces.* Toronto: Key Porter.

Humphreys, S. C. 1969. "History , Economics, and Anthropology: The Work of Karl Polanyi," *History and Theory,* Vol. VIII, No. 2, pp. 165-212.

Jameson, Fredric. 1991. *Postmodernism, Or The Cultural Logic of Late Capitalism.* Durham: Duke University Press.

Jameson, Fredric. 1990. "Postmodernism and Consumer Society" In *Postmodernism and its Discontents.* [ed. E. Ann Kaplan] New York: Verso, pp. 13-29.

Johnson. R. J. 1989. *Environmental Problems: Nature, Economy and the State.* New York: Belhaven Press.

Kaplan, E. Ann. 1991. *Postmodernism and its Discontents.* New York: Verso.

Keats, John. 1973. "Hyperion" In *Romantic Poetry and Prose* [eds. Harold Bloom and Lional Trilling] New York: Oxford University Press, pp. 543-556.

Klinck, Dennis R. 1991. "Tracing a Trace: The Identity of Money in a Legal Doctrine." in *Semiotica,* Vol. 83, Nos. 1/2, pp. 1-31.

Kosik, Karel. 1976. The Dialectics of the Concrete. Boston: D. Reidel Publishing.

Lasch, Christopher. 1991. *True and Only Heaven: Progress and Its Critics.* New York: Norton.

Leiss, William. 1972. *Domination of Nature.* New York: George Braziller.

Leonard, H. Jeffrey [ed.] 1985. *Divesting Nature's Capital.* New York: Holmes and Meier.

Lewis, Martin. 1992. *Green Delusions: An Environmentalist Critique of Radical Environmentalism.* Durham: Duke University Press.

Lewis, Martin W. 1992b. "The Green Threat to Nature." *Harper's Magazine* Nov. p. 26-32.

Livingston, John. 1973. *One Cosmic Instant: A Natural History of Human Arrogance.* Toronto: McClelland and Stewart.

Livingston, John. 1981. *The Fallacy of Wildlife Conservation.* Toronto: McClelland and Stewart.

Livingston, John. 1992. "Les implications ecologiques du caractere imperatif de l'immortalite de l'homme" *Frontieres* vol. 5 nu. 2 Automne.

Livingston, John. 1994. *Rogue Primate: An exploration of human domestication.* Toronto: Key-Porter.

Lukacs, Georg. 1974. *Soul and Form.* London: Merlin Press.

Mackay, Charles. 1980. *Extraordinary Popular Delusions and the Madness of Crowds.* New York: Harmony.

MacNeill, J., P. Winsemius, and T. Yakushiji. 1991. *Beyond Interdependence*. New York: Oxford University Press.

Malinowski, B. 1920. "Kula: The Circulating Exchange of Valuables in the Archipelagoes of Eastern New Guinea," *Man*, No. 51, pp. 97-105.

Marglin, Stephen and Frederique Apffel Marglin. [eds.] 1990. *Dominating Knowledge*. Oxford: Clarendon.

Marx, Karl. 1959. *Capital*. Moscow: Foreign Languages Publishing.

McCay, B. & Acheson, J. [eds.] 1987. *The Question of the Commons*. Tuscon: University of Arizona Press.

McEvoy, Arthur. 1987. "Toward an Interactive Theory of Nature and Culture: Ecology, Production, and Cognition in the California Fishing Industry," *Environmental Review*, Vol. 11, No. 4, pp. 289-305.

McKibben, Bill. 1989. *The End Of Nature*. New York: Random House.

Meeker, Joseph. 1980. *The Comedy of Survival*. Los Angeles: Guild of Tutors Press.

Merchant, Carolyn. 1980. *The Death of Nature: Women, Ecology and the Scientific Revolution*. New York: Harper and Row.

Miller, Daniel. 1987. *Material Culture and Mass Consumption*. Cambridge: Basil Blackwell.

Morse, David. 1981. *Perspectives on Romanticism*. London: Macmillan.

Nelson, Benjamin. 1969. *The Idea of Usury*. Chicago: University of Chicago Press.

Newlyn, W. T. and R. P. Bottle. 1978. *Theory of Money*. Oxford: Clarendon Press.

Niehans, Jurg. 1978. *The Theory Money*. Baltimore: The Johns Hopkins University Press.

Oelschlaeger, Max. 1991. *The Idea of Wilderness*. New Haven: YaleUniversity Press.

Ollman, B. and J. Birnbaum. 1990. *The United States Constitution*. New York: New York University Press.

Ong, Aihwa. 1987. *Spirits of Resistance and Capitalist Discipline*. Albany: State University of New York Press.

Pearson, Harry. 1957. "The Economy Has No Surplus: Critique of a Theory of Development." in *Trade and Market in the Early Empires*. [eds. by Karl Polanyi, Conrad M. Arensberg, and Harry W. Pearson]. New York: The Free Press. pp. 320-339.

Polanyi, Karl. 1947. "Our Obselete Market Mentality," in *Commentary*, Vol. 3, No. 2, pp. 109-117.

Polanyi, Karl. 1957. *The Great Transformation*. Boston: Beacon Press.

Polanyi, Karl. 1968. *Primitive, Archaic and Modern Economies*. [George Dalton ed.] New York: Doubleday.

Porter, Roy. 1982. *English Society in the Eighteenth Century*. Harmondsworth: Penguin.

Postner, Mark. 1984. *Foucault, Marxism and History: Modes of Production versus Modes of Information*. Cambridge: Polity Press.

Repetto, Robert. 1986. *World Enough and Time: Successful Strategies for Resource Management*. New Haven: Yale University Press.

Rogers, Ray. 1991. *Conservation and Development: The Case of Canada's East Coast Fishery*. Unpublished Masters Thesis, Faculty of Environmental Studies, York University, North York.

Rogers, Ray. 1993. "Should Shelburne County Claim the Fish Stocks?" *The Shelburne Coastguard Newspaper*, July 25, p. 9.

Rorty, Richard. 1979. *Philosophy and The Mirror of Nature*. Princeton: Princeton Univ. Press.

Rosen, Jay. 1992. "Playing the Primary Chords." *Harper's Magazine*. March p. 22-26.

Rostein, Abraham and Colin Duncan. 1991. "For a Second Economy" in *The New Era of Global Competition* [eds. D. Drache and M. Gertler] Montreal and Kingston: McGill-Queens University Press, pp. 415-434.

Sachs, Wolfgang. 1992. *The Development Dictionary*. London: Zed Books.

Sahlins, Marshall. 1972. *Stone Age Economics.* Chicago: Aldine.Sante, Luc. 1992.

"The Possessed," *New York Review*, November 12, p. 23.

Sayers, R. S. 1976. *The Bank of England 1891-1944.* New York: Cambridge University Press.

Scientific American. 1990. *Managing Planet Earth.* New York: Freeman.

Sessions, George and William Duvall. 1985. *Deep Ecology.* Salt Lake City: Peregrine Smith.

Shakespeare, William. 1957. *King Lear.* Toronto: Washington Square Press.

Shepard, Paul. 1973. *The Tender Carnivore and the Sacred Game.* New York: Scribners.

Shepard, Paul. 1967. *Man in the Landscape: A Historic View of the Aesthetics of Nature.* New York: Alfred A. Knopf.

Shepard, Paul. 1982. *Nature and Madness.* San Francisco: Sierra Club.

Shiva, Vandana. 1989. *Staying Alive.* London: Zed Books.

Short, John Rennie. 1991. *Imagined Country.* New York: Routledge.

Simmel, Georg. 1990. *The Philosophy of Money.* New York: Routledge.

Smith, Adam. 1961. *The Wealth of Nations.* Indianapolis: Bobbs-Merrill.

Snyder, Gary. 1990. *The Practice of the Wild.* San Francisco: North Point Press.

Spufford, Peter. 1988. *Money and its Use in Medieval Europe.* New York: Cambridge University Press.

Taussig, Michael. 1980. *The Devil and Commodity Fetishism in South America.* Chapel Hill: Univ. of North Carolina Press.

Tawney, R. H. 1982. *The Aquisitive Society.* Brighton: Wheatsheaf.

Taylor, Charles. 1992. *The Ethics of Authenticity.* Cambridge: Harvard University Press.

Temple, Dominique. 1988. "Economicide" *INTERculture*, Vol. 98, pp. 3-35.

Thomas, K. 1983. *Man and the Natural World: A History of the Modern Sensibility.* New York: Pantheon.

Thompson, E. P. 1980. *The Making of the English Working Class.* London: Penguin.

Thoreau, Henry, D. 1962. *Walden.* New York: Houghton Mifflin.

Tillyard, E. M. W. 1942. *The Elizabethan World Picture.* New York: Vintage.

United Nations Environment Programme, International Union for Conservation of Nature and Natural Resources, and World Wide Fund for Nature. 1980. *World Conservation Strategy.* Gland, Switzerland.

Weber, Max. 1958. *The Protestant Ethic and the Spirit of Capitalism.* New York: Scribners.

G. Wheeler. 1990. "Hibernia Blues," *Now Magazine,* Vol. 10, No. 4, Sept. 27-Oct.3, pp. 11-14.

Williams, Raymond. 1982. *Culture and Society.* London: Hogarth Press.

Williams, Raymond. 1980. *Problems in Materialism and Culture.* New York: Verso.

Wilson, Peter J. 1988. *The Domestication of the Human Species.* New Haven: Yale University Press.

Wolin, Sheldon. 1990. "The People's Two Bodies: The Declaration and the Constitution," in *The United States Constitution,* edited by B. Ollman and J. Birnbaum. New York: New York University Press pp. 131-137.

Wood, Ellen Meiksins. 1991. *The Pristine Culture of Capitalism.* New York: Verso.

Wooton, David. 1986. *Divine Right and Democracy.* Harmondsworth: Penguin.

World Commission on Environment and Development. 1987. *Our Common Future.* New York: Oxford University Press.

The World Conservation Union, United Nations Environment Programme, and

World Wide Fund For Nature. 1991. *Caring For The Earth.* Gland, Switzerland.

Worster, Donald. 1977. *Nature's Economy.* New York: Cambridge University Press.

Wrigley, E. A. 1988. *Continuity, Chance and Change.* New York: Cambridge University Press.

INDEX

Appadurai, 55
Barber, 93
Bateson, 91
Baudrillard, 9, 13, 17, 21, 25, 62, 77, 78, 79, 80, 81, 82, 83, 84, 85, 86, 91, 94, 130, 141, 168
Benjamin, 21, 22, 90, 99, 146
Birch, 17, 82
Blake, 125
Bohannan, 48
Brenner, 102, 103
Brown, 8, 9, 25, 50, 51, 52, 53, 54, 55, 84, 85, 86, 91, 106, 113, 115, 158, 168
Canada's East Coast fishery, 1, 4, 6, 7, 8, 56
Capital, 13, 15, 29, 35, 49, 66, 67, 68, 142, 167, 170
capitalism, 1, 9, 10, 15, 19, 20, 24, 27, 53, 65, 66, 67, 68, 69, 78, 83, 84, 90, 99, 101, 103, 107, 118, 128, 141, 146, 149, 151, 154, 155, 157, 159, 160, 167, 170
Clark, 143
commodity fetishism, 68, 70, 85
conservation, 1, 2, 3, 4, 5, 6, 13, 15, 18, 24, 63, 71, 85, 90, 100, 107, 135, 136, 137, 138, 139, 140, 142, 143, 144, 147, 155, 156, 158, 161, 168, 169, 170, 172, 173
Costanza, 139
crisis of modernity, 1, 2, 8, 9, 10, 18, 19, 21, 82, 85, 100, 138, 150, 152, 155, 156, 167, 168, 171
Crump, 26
Daly, 139, 149, 155
de Coppet, 27
death of the subject, 18, 127, 129, 130, 137
development, 1, 2, 3, 4, 7, 16, 24, 25, 31, 36, 40, 42, 43, 48, 49, 50, 51, 55, 67, 71, 79, 83, 85, 91, 102, 103, 116, 119, 128, 129, 136, 137, 138, 139, 140, 141, 142, 143, 144, 149, 154, 155, 159, 162, 166, 169, 170, 172
disembedded, 9, 20, 22, 46, 58, 130
Disembeddedness, 20
Eagleton, 103, 108
Ehrenfeld, 14
embedded, 3, 20, 21, 22, 45, 46, 51, 56, 57, 58, 62, 82, 94, 102, 106, 110, 115, 126, 168, 173
embedded relations, 3, 56, 62, 82, 115, 173
Embeddedness, 20
emergent form, 21, 90, 99, 136, 137, 138, 146
environmental crisis, 1, 2, 9, 10, 16, 18, 19, 50, 74, 109, 135, 136, 138, 140, 146, 152, 169, 171
Evernden, iv, 16, 130, 161, 162
Faust, 52, 150
Foreman, 140, 152, 154
Foucault, 108, 116, 122
Franklin, 64, 65
Freud, 50, 51
global management, 1, 4, 5, 7
globalisation, 7, 9, 10, 16, 90, 99, 137, 146, 14, 150, 158
Goldmann, 98, 100, 120
Goux, 9, 14, 15, 21, 24, 31,62, 70, 71, 72, 73, 74, 75, 76, 77, 78, 79, 84, 85, 86, 92, 95, 130, 169
Grier, 139
Guggenheim, 25
Heilbroner, 66, 136
Hill, 102, 104, 118

Household, 45
Hulme, 119
Industrial Revolution, 9, 90, 99, 116, 117, 118, 120, 122, 146
Jameson, 68, 69, 127, 128
Keats, 90, 99, 121, 173
King Lear, 99, 102, 107, 111, 121, 124
Kosik, 135
Leonard, 142
Lewis, 8, 10, 126, 135, 137, 144, 145, 146, 147, 148, 149, 150, 151, 152, 154, 155, 156, 157, 158, 159, 160, 161, 162, 163, 168, 169, 170, 171, 172
Livingston, vii, 2, 10, 16, 17, 92, 93, 94, 95, 97, 98, 100, 129, 152, 157, 169, 171
Luther, 53, 158
MacNeill, 139, 142
Malinowski, 27
Marglin, 140, 141, 142
markets, 1, 2, 7, 13, 16, 45, 46, 47, 48, 49, 50, 56, 58, 90, 101, 103, 158, 161
Marx, 2, 8, 13, 25, 26, 29, 30, 31, 32, 33, 34, 35, 36, 37, 40, 41, 42, 49, 56, 57, 63, 65, 66, 67, 70, 73, 77, 78, 79, 80, 81, 86, 120, 159, 169, 170
Miller, 62, 63, 64, 73
Morse, 100, 116
Mowat, 5
Nelson, 65, 106
Niehans, 26
Oelschlaeger, 126, 127
Ollman, 126
Ong, 24, 44
Our Common Future, 3, 4, 139
Pearson, 49
Polanyi, 2, 8, 17, 18, 20, 25, 28, 43, 44, 45, 46, 47, 48, 50, 56, 57, 63, 73, 80, 81, 86, 91, 94, 117, 169, 174
Poster, 78
Protestant, 52, 53, 102, 106
Protestantism, 41, 42
reciprocity, 2, 17, 45, 46, 52, 81, 82, 86, 94
redistribution, 2, 45, 46, 86
reification, 41, 57, 68, 70, 85
Repetto, 143, 149
residual forms, 8, 21, 22, 89, 90, 99, 101, 127, 146, 153
Romanticism, 100, 118, 120, 124, 172
Rosen, 129
Rotstein, vi, 47, 56
Sachs, 5, 141, 142, 162
Sahlins, 20, 28
Sante, 130
Shakespeare, 13, 90, 99, 103, 104, 106, 173
Shepard, 17, 168
Simmel, 8, 14, 25, 36, 38, 39, 40, 41, 42, 43, 57, 70, 86, 89, 91, 99, 120, 156, 169, 170
Smith, 25, 26, 36
social transformation, 9, 19, 22, 24, 99, 103, 105, 106, 115, 118, 152
Spivak, 1
Strong, 139
sustainability, 1, 3, 135, 136, 138, 139, 140, 144, 170, 171
Taussig, 12, 18, 21, 22, 28, 53, 66, 67, 68, 96, 97, 98, 170
Taylor, 10, 167, 169
Temple, 46
Thompson, 118, 125
Tragedy, 107, 115, 116, 120, 122, 125, 171, 172
Turner, 149
Williams 21, 89, 96, 98, 100, 119, 120, 150
Winstanley 105, 106, 113, 114
Wolin 126
Wood vii, 101, 103
Wooten, 105
Wordsworth, 125
Wrigley, 116, 117

Also published by

by MURRAY BOOKCHIN

THE PHILOSOPHY OF SOCIAL ECOLOGY

second edition, revised

"A useful corrective to simplistic thinking about the human predicament."

Canadian Book Review Annual

"Bookchin expands upon the concept of natural evolution and delivers it from the trap of mechanistic thinking that obscures its potential for rational understanding of the natural world."

Imprint

245 pages
Paperback ISBN: 1-551640-18-X $19.99
Hardcover ISBN: 1-551640-19-8 $38.99

THE ECOLOGY OF FREEDOM

The Emergence and Dissolution of Hierarchy

revised edition

...a confirmation of his [Bookchin's] status as a penetrating critic not only of the ways in which humankind is destroying itself, but of the ethical imperative to live a better life.

The Village Voice

Elegantly written, and recommended for a wide audience.

Library Journal

395 pages, index
Paperback ISBN: 0-921689-72-1 $19.99
Hardcover ISBN: 0-921689-73-X $38.99

REMAKING SOCIETY

To anyone new to his work, *Remaking Society* provides a clear synthesis of Bookchin's ideas.

...an intellectual tour de force...the first synthesis of the spirit, logics, and goals of the European "Green Movement" available in English.

Choice

208 pages
Paperback ISBN: 0-921689-02-0 $18.99
Hardcover ISBN: 0-921689-03-9 $37.99

TOWARD AN ECOLOGICAL SOCIETY

3rd printing

Bookchin's great virtue is that he constantly relates his theories to society as it is.

George Woodcock

Bookchin is capable of penetrating, finely indignant historical analysis. Another stimulating collection.

In These Times

A provocative work that gives abundant evidence of its author's position at the centre of the debate.

Telos

315 pages
Paperback ISBN: 0-919618-98-7 $18.99
Hardcover ISBN: 0-919618-99-5 $37.99

POLITICAL ECOLOGY

Beyond Environmentalism

Dimitrios Roussopoulos

Examining the perspective offered by various components of political ecology, this book presents an overview of its origins as well as its social and cultural causes, and summarizes the differences, and similarities, between political ecology and social ecology.

180 pages
Paperback ISBN: 1-895431-80-8 $15.99
Hardcover ISBN: 1-895431-81-6 $34.99

BALANCE: Art and Nature

John K. Grande

In this timely book, written with headlong urgency, John Grande encourages us to rethink what it means to be an artist in a time of global eco-crisis. His valuable perspective would liberate art from the stultifying value-system of consumer capitalism and bring it into a more coherent relationship with nature's processes. It is inspiring to read about artists who are using their creativity for constructive thought and action.
Suzi Gablik

205 pages, photographs
Paperback 1-551640-06-6 $19.99
Hardcover 1-551640-07-4 $38.99

HUMANITY, SOCIETY, AND COMMITMENT

On Karl Polanyi

Kenneth McRobbie, ed.

Karl Polanyi's lesser-known writings, his correspondence, teaching, and views on the future of socialism are brought together for the first time in this exciting and timely collection. The contributors include: Michele Cangiani, Marguerite Mendell, Marco Eller Vainscher, Endre Nagy, Kari Polanyi-Levitt, Daniel Fusfeld, Abraham Rotstein, and Kenneth McRobbie.

210 pages
Paperback ISBN: 1-895431-84-0 $19.99
Hardcover ISBN: 1-895431-85-9 $38.99

ARTFUL PRACTICES

The Political Economy of Everyday Life

Henri Lustiger-Thaler and Daniel Salée, eds.

The social, economic and political institutions we have lived with since the post-war settlement are undergoing transformations that are deeply challenging notions of universality, the nature of the market, and the role of citizenship. As citizens are forced into fierce competition over scarce resources, the terms of the social contract are being renegotiated. The artful practices of citizens coping with these massive changes point to a new political economy of everyday life.

200 pages
Paperback ISBN: 1-895431-92-1 $19.99
Hardcover ISBN: 1-895431-93-X $38.99

Printed by the workers of
Les Éditions Marquis
Montmagny, Québec
for Black Rose Books Ltd.